Women and Climate Change in Bangladesh

Bangladesh is by no means a high emitter of carbon, but it is nevertheless one of the countries most critically affected. There is a significant risk of damage to lives and livelihoods due to climate change in the form of cyclones, flooding and storm surges, and slow-onset impacts such as droughts, sea level rises and river basin erosion. Moreover, Bangladeshis are especially vulnerable, as a high proportion of people live in extreme poverty. This book assesses the impact of climate change in Bangladesh and presents the findings of a three-year, in-depth study undertaken at village level in different districts of the country. It examines national policies, contrasting them with what is actually happening at village level. It outlines the impact of climate change on livelihood strategies and health, and focuses particularly on the impact on gender relations, showing that, although women have a significant role to play in helping communities cope with the effects of climate change, cultural customs and practices often work against this. The book argues for, and puts forward policy proposals for, recognizing women's active contribution and supporting gender equality as a critical strategy in global adaptation to climate challenges.

Margaret Alston is Professor in the Department of Social Work at Monash University, Australia.

ASAA Women in Asia Series

Mukkuvar Women
Gender, hegemony and capitalist transformation in a South Indian fishing community
Kalpana Ram 1991

A World of Difference
Islam and gender hierarchy in Turkey
Julie Marcus 1992

Purity and Communal Boundaries
Women and social change in a Bangladeshi village
Santi Rozario 1992

Madonnas and Martyrs
Militarism and violence in the Philippines
Anne-Marie Hilsdon 1995

Masters and Managers
A study of gender relations in urban Java
Norma Sullivan 1995

Matriliny and Modernity
Sexual politics and social change in rural Malaysia
Maila Stivens 1995

Intimate Knowledge
Women and their health in North-east Thailand
Andrea Whittaker 2000

Women in Asia
Tradition, modernity and globalisation
Louise Edwards and Mina Roces (eds) 2000

Violence against Women in Asian Societies
Gender inequality and technologies of violence
Lenore Manderson and Linda Rae Bennett (eds) 2003

Women's Employment in Japan
The experience of part-time workers
Kaye Broadbent 2003

Chinese Women Living and Working
Anne McLaren (ed.) 2004

Abortion, Sin and the State in Thailand
Andrea Whittaker 2004

Sexual Violence and the Law in Japan
Catherine Burns 2004

Women, Islam and Modernity
Single women, sexuality and reproductive health in contemporary Indonesia
Linda Rae Bennett 2005

The Women's Movement in Post-Colonial Indonesia
Elizabeth Martyn 2005

Women and Work in Indonesia
Michele Ford and Lyn Parker (eds) 2008

Women and Union Activism in Asia
Kaye Broadbent and Michele Ford (eds) 2008

Gender, Islam, and Democracy in Indonesia
Kathryn Robinson 2008

Sex, Love and Feminism in the Asia Pacific
A cross-cultural study of young people's attitudes
Chilla Bulbeck 2008

Gender, State and Social Power
Divorce in contemporary Indonesia
Kate O'Shaughnessy 2008

Gender, Household, and State in Post-Revolutionary Vietnam
Jayne Werner 2008

Young Women in Japan
Transitions to adulthood
Kaori Okano 2009

Women, Islam and Everyday Life
Renegotiating polygamy in Indonesia
Nina Nurmila 2009

Feminist Movements in Contemporary Japan
Laura Dales 2009

Gender and Labour in Korea and Japan
Sexing class
Ruth Barraclough and Elyssa Faison (eds) 2009

Gender Diversity in Indonesia
Sexuality, Islam and queer selves
Sharyn Graham Davies 2010

New Women in Colonial Korea
A sourcebook
Hyaeweol Choi 2012

Women Writers in Postsocialist China
Kay Schaffer and Xianlin Song 2013

Domestic Violence in Asia
Globalization, gender and Islam in the Maldives
Emma Fulu 2014

Gender and Power in Indonesian Islam
Leaders, feminists, Sufis and pesantren selves, edited
Bianca J. Smith and Mark Woodward 2014

Practicing Feminism in South Korea
The women's movement against sexual violence
Kyungja Jung 2014

The Korean Women's Movement and the State
Bargaining for change
Seung-kyung Kim 2014

Gender, Nation and State in Modern Japan
Andrea Germer, Vera Mackie and Ulrike Wöhr (eds) 2014

Women and Sex Work in Cambodia
Blood, sweat and tears
Larissa Sandy 2015

Growing up Female in Multi-Ethnic Malaysia
Cynthia Joseph2015

Women, Sexual Violence and the Indonesian Killings of 1965–66
Annie Pohlman 2015

Love and Marriage in Globalising China
Wang Pan 2015

Women and Climate Change in Bangladesh
Margaret Alston 2015

Women and Politics in Contemporary Japan
Emma Dalton 2015

Trafficking Women in Korea
Filipina migrant entertainers
Sallie Yea 2015

Women and Sharia Law in Northern Indonesia
Dina Afrianty 2015

China's Leftover Women
Late marriage among professional women and its consequences
Sandy To 2015

Women and Climate Change in Bangladesh

Margaret Alston

Routledge
Taylor & Francis Group
LONDON AND NEW YORK

First published 2015
by Routledge
2 Park Square, Milton Park, Abingdon, Oxfordshire OX14 4RN
711 Third Avenue, New York, NY 10017

Routledge is an imprint of the Taylor & Francis Group, an informa business

First issued in paperback 2017

British Library Cataloguing in Publication Data
A catalogue record for this book is available from the British Library

Library of Congress Cataloging in Publication Data
Women and climate change in Bangladesh / Margaret Alston.
pages cm. – (ASAA women in Asia series)
Includes bibliographical references and index.
1. Women and the environment–Bangladesh. 2. Human beings–Effect of climate on–Bangladesh. 3. Climatic changes–Social aspects–Bangladesh. 4. Climatic changes–Economic aspects–Bangladesh. 5. Women–Bangladesh–Social conditions. 6. Women–Bangladesh–Economic conditions. 7. Community development–Bangladesh. I. Title.
HQ1745.6.A874 2015
305.4095492–dc23
2014027961

ISBN: 978-1-138-02604-9 (hbk)
ISBN: 978-0-8153-6173-2 (pbk)

Typeset in Times New Roman
by Taylor & Francis Books

Contents

Illustrations

Figures

Tables

Boxes

Acknowledgement

This book owes everything to the people of Bangladesh who took part in this research, giving their time and energies to ensure that their voices will be heard across the world. As one woman in a rural Bangladeshi village in the Bay of Bengal asked me directly – 'Will our voices be heard?' Through this work I hope to not only ensure that her voice is heard but that the world listens.

To the government and non-governmental organisation (NGO) representatives in Dhaka who participated, attended workshops and gave valuable feedback, my sincere thanks.

This research was made possible by an anonymous donation to the Oxfam-Monash Partnership and the members of the research team mentioned in these acknowledgements. I especially want to thank Dr Kerri Whittenbury, Alex Haynes and Naomi Godden from Monash University; Tania Cass, Chrisanta Muli and colleagues at Oxfam Australia; MD Badi Akhter, Julia Ahmed, Hasina Henna and Sonia Sarder at Oxfam Great Britain; staff of Oxfam Great Britain's local development partners, Jagonari in the Barguna District, GUK in the Gaibanda District and Shushilan, Satkira District, for their invaluable assistance with site visits and the complex logistics associated with conducting research in isolated areas; Fahmid Al Zaid, Kazi Nimery, Nusrat Sharmin Diya and Mahfuza Akhter Mala and their Professor Hasan Al Shafie, who assisted in collecting the data, translating and transcribing this valuable material; Ms. Nazme Sabina, supported by Bipasha Dutta and Shusmita Khan, from Eminence Consultants who conducted the survey research; and Craig Poynter from Charles Sturt University for working hard to produce the maps – thank you.

Without the Partnership's foresight in choosing this project, this work would never have been possible. Through the Oxfam network we have had unprecedented access across the country, working closely with Oxfam Australia and Oxfam Great Britain colleagues and their partner organisations within the country.

To my colleagues at Monash University, Dr Kerri Whittenbury and Alex Haynes, a special thanks for their ongoing and vital support during the extraordinary three-year period of this research, their data gathering assistance, company on field visits and extensive wisdom – thank you. Finally to my husband for his wit, wisdom and support – my sincere thanks. The views expressed in this book are my own and do not necessarily reflect those of the Oxfam-Monash partnership or of Oxfam or its affiliates.

Series Editor's Foreword

The contributions of women to the social, political and economic transformations occurring in the Asian region are legion. Women have served as leaders of nations, communities, workplaces, activist groups and families. Asian women have joined with others to participate in fomenting change at micro and macro levels. They have been both agents and targets of national and international interventions in social policy. In the performance of these myriad roles women have forged new and modern gendered identities that are recognisably global and local. Their experiences are rich, diverse and instructive. The books in this series testify to the central role women play in creating the new Asia and re-creating Asian womanhood. Moreover, these books reveal the resilience and inventiveness of women around the Asian region in the face of entrenched and evolving patriarchal social norms.

Scholars publishing in this series demonstrate a commitment to promoting the productive conversation between Gender Studies and Asian Studies. The need to understand the diversity of experiences of femininity and womanhood around the world increases inexorably as globalisation proceeds apace. Lessons from the experiences of Asian women present us with fresh opportunities for building new possibilities for women's progress the world over.

The Asian Studies Association of Australia (ASAA) sponsors this publication series as part of its on-going commitment to promoting knowledge about women in Asia. In particular, the ASAA Women's Forum provides the intellectual vigour and enthusiasm that maintains the Women in Asia Series (WIAS). The aim of the series, since its inception in 1990, is to promote knowledge about women in Asia to both academic and general audiences. To this end, WIAS books draw on a wide range of disciplines including anthropology, sociology, political science, cultural studies, media studies, literature, and history. The series prides itself on being an outlet for cutting edge research conducted by recent PhD graduates and postdoctoral fellows from throughout the region.

The Series could not function without the generous professional advice provided by many anonymous readers. Moreover, the wise counsel provided by Peter Sowden at Routledge is invaluable. WIAS, its authors and the ASAA are very grateful to these people for their expert work.

Lenore Lyons (The University of Sydney)
Series Editor

1 Introducing climate change and its impacts on the people of Bangladesh

Bangladesh is a predominantly Islamic country with a population approaching 160 million living in one of the most densely populated regions of the world. It is a low-lying, richly fertile country criss-crossed by majestic rivers, belying its extreme vulnerability to climate change and variable weather patterns. Over the last two decades it has experienced multiple, catastrophic climatic events including droughts, river erosion, increasingly intense cyclonic activity, storm surges, salt water inundation, hotter, drier summers and escalating challenges to food production cycles and water security. The impacts of these events on the people of Bangladesh are wide ranging and affect every facet of their lives. This book focuses on the impacts of these events on the people who live in this beautiful, environmentally sensitive country, a country amongst the most vulnerable on the planet to climate change. The challenge of this work is to bring the voices of those most critically affected to the world's attention and to make them a significant part of the global conversation on climate actions.

A particular focus of this work is to illuminate the lives of the women of Bangladesh. The volatile historical record of this relatively new country pays homage to the women who fought, suffered and died during the War of Liberation in the 1970s. Extraordinary advances in women's status have occurred in recent decades. There is now near parity in primary education access for girls and boys, and women occupy 20 per cent of parliamentary positions. However, these advances hide the significant disadvantages that they face. Women own very little of their country's land mass despite their very active participation in agricultural production and labour. Further, their recognition in the nation's constitution as equal citizens masks the overriding influence of the Islamic religion and attendant personal, or Sharia, laws on women's lives.

The emergence of a strong women's movement during the 1970s and 1980s established women's political voice and activism and laid the foundation for women's human rights to be recognized and valued. Yet, as this book demonstrates, women's lives are dependent on education and circumstances, on power and oppression, on poverty and livelihood capacity and on their links to land, the environment, information, training and infrastructure. Many women in Bangladesh go hungry to support their children and most work long hours into old age to ensure the survival of their families. There is a strong and enduring spirit amongst the women who took part in this research and a hope for the future. In presenting their views, it is my hope that global cooperation on climate change might present an opportunity to build the capacity and empower the women of Bangladesh to be major leaders in an uncertain future.

The world watches

Climate change is one of the global community's most pressing challenges. It is a problem of extraordinary proportions, with differential impacts within and between countries exacerbating global inequalities and foreshadowing immense change. Speaking at the 18th Conference of the Parties (COP) in Doha in December 2012, the United Nations Secretary-General, Ban Ki-Moon, described it as a major threat to humanity:

> Let us be under no illusion. This is a crisis. A threat to us all … It is an existential challenge for the whole human race – our way of life, our plans for the future. We must take ownership. We, collectively, are the problem. Then we should have the solutions.
>
> (United Nations News Centre 2012)

Despite his exhortation, the global community is struggling to develop cooperative and sustainable solutions. While the Conference of the Parties meeting in Warsaw in 2013 resulted in a commitment to a universal climate agreement, there appears to be limited cooperation on immediate global actions. This leaves the most affected countries like Bangladesh to deal with the overwhelming consequences that are occurring now. Ironically the most vulnerable countries are least likely to be responsible for the earth's changing climate, and those with the greatest capacity to act are failing to do so. How do we best address this challenge to ensure that we preserve and sustain the planet for future generations? How do we deepen the climate change conversation to include socially just solutions that require intense and open global dialogue? And how do we ensure strategic actions to assist the most vulnerable people and communities?

The process of developing the solutions demanded by the Secretary-General is destabilized by a number of factors, not least a conservative backlash against global cooperation and a retreat by many countries to a position that is focused on their own best interests. Also culpable is the language surrounding 'climate change', the nature of climate crises, the unevenness of impacts and the varied vulnerability of diverse groups – a language driven by a neoliberal economic agenda. What is missing from the conversation are the voices of people most directly and devastatingly affected. These have been overshadowed by the intense necessity to come to grips with the scale of the changes, the scientific analysis of causes and effects and an assessment of economic outcomes. It is arguable that this has led to climate change being cast as an environmental problem requiring economic solutions. The 2006 Stern Review, for example, described it as the 'greatest and widest ranging market failure ever seen' (Smith 2006; Stern 2006).

Hence there is significant attention in many countries to carbon pricing, emissions-trading schemes and technological solutions. Australia, for example, opted for a contentious Carbon Emissions Trading Scheme in 2012, a form of taxation designed to reduce carbon emissions and encourage alternative energy sources. Arguably this became a significant factor in the 2013 political campaign that resulted in the election of a very conservative government committed to throwing out the scheme and its reframing as an unwanted tax. The embedded economic language surrounding discussions of climate change moves the focus of discussions away from people and towards more divisive economic measures. Further, the scepticism of many in the global community about the veracity of climate science has led to a polarized debate as to whether scientific proof is strong enough to view climate variability as more than an aberration – an outlier in a pattern that will inevitably return to normal.

Nonetheless both the move to economic responses and the need to prove a case for climate change lead to ever more precise science and technological developments designed to aid mitigation initiatives and to find adaptive responses. This top-down scientific certainty in dealing with climate change has led to a form of global managerialism (Tschakert *et al.* 2011; Alston 2013) that presupposes that all people can adapt with determined actions, the right technology and adequate funding to enhance their capacity.

What we lose in this approach is an understanding of the differential impact within and between countries, the significant differences in the capacity to adapt, the messiness of everyday life, the complexity of poverty and disadvantage, and the differential experiences of women and men, the old and the young, and the rural and urban enduring similar circumstances with very different resources, capacity and resilience. Global managerialism and its fixation on linear, apolitical solutions reduces attention to socially just responses, to the human rights of those most affected and to the capacity of vulnerable individuals and communities to act in logical and deterministic ways.

This book goes much further than most common discussions of vulnerable populations. It presents evidence that women and men are differentially impacted and that the impacts on women and girls are markedly more extreme. Climate change and its fundamental challenges are presented from the position of people in a country with high levels of poverty and significant climate challenges. I focus on those in rural villages and, in particular, on rural women and girls who are experiencing 'climate change' in challenging ways. These research findings emerge from three years of study across three diverse regions of the country facing quite different climatic impacts. In presenting these findings this book brings forth the face of climate change, adding a social perspective to developing knowledge in this area, and represents climate change threats as far more than environmental and economic catastrophes. This work demonstrates how climate change is causing major social upheaval, facilitating diverse coping strategies (not all of them just or sustainable) and creating ongoing changes at household levels that are reshaping lives and communities across the globe.

The people represented in this book are responding with limited means to very complex and diverse climatic and environmental challenges. It is in the representation of their voices and understandings that we come to see climate change as a social phenomenon – one with the capacity to challenge food production and water security and people's relationship to land and community. This book destabilizes the notion that climate change adaptation is a linear process requiring ever more precise knowledge and technology. It notes that the critical, knowledge held by women and men at local levels must be part of global solutions. This work shows how traditional customs and values sometimes reinforce unsustainable and socially unjust practices. And thus it notes that climate change has the capacity to reinforce existing gender inequalities in perverse ways at the same time as it holds the potential to challenge and build gender equality. The writer views climate challenges as a critical space to explore gender actions and strategies and to critique previously taken-for-granted, seemingly immutable customs and practices. It is therefore argued that climate change be considered as a site with the revolutionary potential to address intransigent gender inequalities.

Climate change

Climate change has become a hold-all basket for all things related to weather variability, leaving people largely uncertain about its nature and complexity. The widespread tendency amongst high-emitting countries to treat attention to climate change as a tax-based emissions reduction scheme not only creates confusion, but also reduces general understanding. Issues such as poverty, health, welfare and differential gender impacts are missing from international discussions.

Climate change is defined more precisely in the United Nations environment as 'a change of climate which is attributed directly or indirectly to human activity that alters the composition of the global atmosphere and which is in addition to natural climate variability observed over comparable time periods' (UNFCCC 2011). This build-up of greenhouse gases has caused rising global temperatures and melting of the polar icecaps, sea level rises, longer and more intense droughts, and increased catastrophic events such as cyclones, floods, mudslides and storm surges.

These changes are having, and will have, direct health effects on people as a result of changes in exposure to extremes of weather (heat waves or cold snaps); increased airborne pollutants and aeroallergens; indirect effects as a result of the transmission of water and airborne infectious diseases such as malaria and cholera; decreased food production yields both on land and in the sea; decreased access to fresh water; and loss of biodiversity (McMichael 2003). Ample evidence emerges from the research findings presented in this book that, in areas experiencing climate challenges, health impacts also include skin diseases, stress and anxiety and gynaecological problems resulting from long periods in waterlogged areas.

International institutional responses to climate change

The United Nations Framework Convention on Climate Change (UNFCCC) launched in 1990 was designed to assist the global community to reduce greenhouse gas emissions. The first meeting of the UNFCCC was informed by the International Panel on Climate Change (IPCC) *First Assessment Report*. The IPCC brings together scientists from across the world to examine evidence, present reports on the global situation and advise on actions. Their fifth report was produced in 2013/2014. The first IPCC report (1991) set a global target of a 20 per cent reduction in emissions. Subsequently this was ratified in the lead-in discussions to the first Earth Summit in 1992.

Each year the Conference of the Parties meetings are held to discuss the UNFCCC and the IPCC reports. While initially there was extraordinary global cooperation to assess the problem and to find global solutions, this has not continued. Annual discussions on reduction targets have resulted in stalemates, frustrations and blocking mechanisms as differing levels of national self-interest and influence play out. The fifteenth COP meeting held in Copenhagen in 2009 appeared to be the best chance for global consensus and cooperation in addressing climate strategies. One hundred and fifteen of the world's leaders came together to develop a binding agreement to limit greenhouse gas emissions and temperature rises, but this, too, failed. The 2013 COP 19 in Warsaw also resulted in much rhetoric and limited actions and a more vocal resistance group, including Australia, working against binding agreements.

Over the last two decades the world has seen the ambitious targets for greenhouse emissions reductions reduced and watered down, first to a target of a 5 per cent reduction agreed at the 1997 Kyoto meeting, and now more recently to ill-defined general emissions targets. In addition, the conversation has shifted to how the global community can limit temperature rises to 2 degrees to avoid further destabilizing future food and water security, and reducing health impacts – goals that at present seem equally unattainable. There is now considerable concern that a lack of global effort will result in a 4 degree rise this century.

One would anticipate significant global attention to this major threat to stability. Yet, at regular COP meetings designed to address this catastrophic challenge to the future of humanity, there has been a lack of consensus on actions and a timidity in language that is at best alarming. It is apparent that countries that are the worst emitters of greenhouse gases are the most powerful economies, and that they have greater capacity in terms of resources and infrastructure to address immediate outcomes of climate-induced disasters and challenges. Thus, countries such as the United States, while not immune to climate disasters as evidenced by Hurricane *Katrina* in 2004 and major flooding in New York in 2012, are resistant to global cooperation on climate actions, fearing that their economies may suffer. At the same time, powerhouse economies in the developing world such as India and China are loathe to support global emissions reduction targets whilst their economies are developing so rapidly. One result of the variable exposure to climate change events, and the differing capacity to respond, is that most countries are unwilling or unable to support a global response.

Those countries at the greatest threat are more typically least resilient. The Pacific Island nations, for example, are subject to sea level rises that may threaten their entire countries. Yet they have more limited capacity to influence world opinion and are less able to adapt to the ongoing impacts of climate events. Already there are moves underway to shift populations from some of the lower islands. Tuvalu Prime Minister, Apisai Ielemia, argued in a speech to the UN General Assembly in 2009:

> We strongly believe that it is the political and moral responsibility of the world … to save small islands and countries like Tuvalu from climate change, and ensure we continue to live in our home islands with long-term security, cultural identity, and fundamental human dignity.
>
> (Oxfam 2009: 34)

Unfortunately, while the likely impacts on Island nations are known by local leaders and acknowledged by global policymakers, there has been limited success in mobilizing actions to secure their future. The intense frustration felt by representatives of these countries resulted in a walk-out of delegates at the Warsaw COP meeting. This book focuses on a country that is extremely fragile and has a limited capacity to respond. Bangladesh is one of the most severely affected by climate change events, both catastrophic and slow-onset, and has an equally uncertain future.

Food and water security

Meanwhile the global hand-wringing and inaction on social aspects of climate change will result in significant consequences in a very short period of time. We are moving into a future where we cannot produce enough food in the right places to sustain an ever-increasing population, predicted to reach 9 billion by 2040. This, together with economic destabilization, austerity measures in many countries, political instability and declining productivity in food production signals a period of increasing risk for those most compromised by changes.

Britain's Chief Scientist, Sir John Beddington, has warned that global warming and temperature rises will result in much greater variability in weather, and therefore food production (Linden 2013), and that food and water insecurity resulting from climate change are 'the elephant in the room' for the global community (Randerson 2008). Estimates are that the number of hungry and malnourished people will rise by more than 10 per cent in the coming decade and will exacerbate poverty among the most vulnerable (FAO 2007a). In a sign that this is escalating, the frequency of food riots and the number of countries affected are increasing.

The International Fund for Agricultural Development (IFAD 2007) notes that most of the world's 1.2 billion poor (two-thirds of whom are women) live in countries where many people do not have access to safe, reliable water. Further, they note that one-third of the world is experiencing some form of water shortage. IFAD suggests that inequities in access to water are increasing globally because of competition for limited resources. The World Health Organization (2012) notes that the 20 per cent of Bangladeshis who live in the Char and Hill areas will increasingly find it difficult to access clean water. Even in the capital, Dhaka, ground water is dwindling and salinity is increasing. Amongst other critical impacts is the effect on women's time – it takes much longer to collect water for the household in many areas. Further, up to 60 per cent of Bangladeshis lack access to safe sanitation, and this has critical impacts on the spread of disease.

The language of climate change

How climate change is understood is critical to the way in which actions and strategies to mitigate its effects are determined. Language is never innocent. The way *climate change* has been constructed to reflect particular views and to prioritize certain solutions is a potent illustration of the power/knowledge intersection articulated by prominent French philosopher and social theorist, Michel Foucault. Foucault argues that the construction of discourses, or the meanings we give to certain issues, are filtered and shaped from a particular standpoint that favours the interests of dominant individuals and nations. Thus, particular 'truths' articulated by the powerful, hold sway (Foucault 1991, 1998: 63). Foucault argues that power circulates in such a way that the powerful shape these truths, and that discourse, knowledge and regimes of truth shape power. Thus power/knowledge is constructed through processes that build, confirm and continually refine views that support and sustain those in powerful positions.

Using this lens it is possible to see that power is a central factor in climate change discussions and negotiations and that power relations have shaped the way *climate change* is constructed, enunciated and addressed. In the development of our understanding of climate change, powerful nations and stakeholders have built a dominant construction or 'truth' that is largely depopulated, disconnected from local realities and certainly ungendered.

Feminist theorists argue that the development of knowledge and the construction of discourses demand that we interrogate whose 'truths' are dominant and for what purpose. As Hawthorne (2001) asks – 'Who is the knower?' Who is constructing the dominant form of understanding of this particular issue? Haraway (1988) had earlier argued that all knowledge represents power rather than any essential truth. She argues for an embodied subjectivity – or situated, embodied knowledge – that denotes and makes visible the truth tellers or knowledge shapers:

> Struggles over what will count as rational accounts of the world are struggles over how to see … I am arguing for the view from a body, always a complex, contradictory, structuring and structured body versus the view from above, from nowhere … the god trick … rational knowledge is a process of ongoing critical interpretation among fields of interpreters … rational knowledge is power-sensitive conversation.
>
> (Haraway 1988: 587–590)

The interpretation of climate change would appear to demonstrate a particular form of power-sensitive conversation, the 'god trick', a construction that largely ignores the particular vulnerability of countries, regions and the women and men within these fragile environments.

Hawthorne (2001) urges that we understand that knowledge exists only in relation to its social location, a view supported by Dorothy Smith (1992: 17) who stresses that we must take into account the context of women's lives – the *everyday/everynight worlds* in our construction of knowledge. Further, by taking account of the local, cultural and environmental circumstances of particular groups we can *resist the universalizing impulse* (Hawthorne 2001: 51) inherent in scientific, undifferentiated 'truths' – in this instance the decontextualized and depopulated understanding of climate change. Thus, as Hawthorne notes, feminist theorists prefer to develop knowledge or *truths* that incorporate consciousness-raising and engagement; strong reflexivity (Harding 1992: 163); decolonizing methodologies (Tuhiwai Smith 1999); a passionate engagement with research questions and a spurning of the view of the researcher as expert. In particular, feminist researchers urge that research and outcomes remain grounded in everyday experience, thereby focusing attention on critical social and gendered impacts and solutions that build resilience. This stands in stark contrast to the universalizing tendency of science that becomes disconnected, captured by worldviews and shaped by those with interests in supporting certain regimes of truth.

Scientists and policymakers are prominent in shaping climate change as an environmental and economic problem demanding scientific, technological and economic solutions. This disembodied and decontextualized understanding allows us to ignore a different construction, one that is populated by women and men, girls and boys, whose lives are constantly under threat and whose power to construct a view of climate threats that includes fear, challenge and disrupted lives is muted.

Indian eco-feminist, Vandana Shiva, notes that the dominant form of scientific reductionism deflects attention from the complexity of destabilized regions and results in an understanding of the world that is limited, undemocratic and harmful. She suggests that scientific reductionism is based on an assumption of homogeneity (1989: 22). In her book, *Earth Democracy* (Shiva

2005), she notes that principles of peace, justice and sustainability can only emerge from a more self-aware acknowledgement of the connection between people and the planet, an awareness that demands not only an understanding of inequalities but also a democratic, inclusive responsibility for preserving the environment. By contrast, the dominant scientific view produces a hierarchy of nation states and private interests, prioritizes the privileges of large-emitting nations at the expense of low-emitting and/or fragile states, and the wealthy over the poor and ultra-poor. In the business of addressing climate change, there is an increasing gap between transnational actors and local-level experiences.

What emerges from this view of climate change is a global managerialist approach that:

- is disconnected from the complexities of the lives of those most affected;
- views the process of adaptation as linear and apolitical;
- directs adaptive responses to local levels in a patchy, largely uncoordinated way;
- fails to interrogate inequalities;
- is gender-blind; and
- presumes that, if enough resources are provided, adaptation is possible by anyone anywhere.

(Tsachkert *et al.* 2011)

This position largely ignores capacity to adapt, the vulnerability of communities and individuals and interventions that might enhance resilience. Hawthorne (2001), drawing on Wangoola (2000), defines multiversity as a system of knowledge generation that challenges the disconnection and abstraction of conventional knowledge generation. She argues that we must build understanding from the experiences of people within their cultural context, enabling dialogue, partnership, collaboration and coherence in the building of solutions and the empowerment of people at local levels. This book outlines the experiences of women and men, girls and boys in one of the most vulnerable and fragile nations in the world. In presenting this information, a more detailed, contextually grounded understanding of climate change is provided.

Gender, vulnerability and resilience

Successful adaptation to climate events is dependent on the resilience of affected people and communities. Resilience is defined by the IPCC (2008) as the ability to absorb changes in a positive way. This is not possible if people feel unsupported, if change appears risky, if traditional ways of living appear safer, and if institutional supports are not forthcoming (Tschakert *et al.* 2011). Resilience in the face of climate events requires coordinated social supports, and reflective transformation of cultural constraints and practices, of livelihood strategies, of resource use, and of institutional structures and functions (Pelling 2011). To move adaptations in a positive, sustainable direction, requires that these actions and transformations be sensitive to gender equality and empowerment strategies for women and girls.

In the context of climate change, vulnerability has been defined as the capacity to 'anticipate, cope with, resist and recover from the impact of a natural hazard' (Kelly and Adjer 2009: 163). Vulnerability commonly refers to factors that in combination produce a lack of capacity to respond effectively, in this case to climate variations and disasters. Those most vulnerable are characterized by gender, poverty, educational disadvantage, reduced service access, lack of employment options and other aspects of socio-economic disadvantage.

The regular reconstitution of gender inequalities results in the greater vulnerability of women in many disaster circumstances. This vulnerability is not caused by climate events – it is exacerbated by them. It results from a much more complex set of circumstances, linking existing gender inequalities such as reduced access to land, resources and decision-making, constraints on freedom of movement, lower income-generation capacity, disproportionate caring roles, traditional cultural customs and gender-blind policies. A more nuanced explanation of these complexities is required in order to open the possibility of greater gender equality, a more equitable distribution of resources and greater resilience in the face of climate challenges.

Enarson (2012: 2) notes the *striking disregard for gender* in the context of disasters despite the evident differences in impacts and outcomes. Gender is defined as the social processes and relations that shape our identity as women or men at a certain time and place. Gender is, as Enarson (2012: 23–26) notes, a marker of difference, identity and inequalities; it is relational – existing only in relation to others; it is socially constructed and plays out at individual and institutional levels; it is reinforced through social practices, and is forever dynamic.

Many countries accept and endorse unequal gender relations, leading to reduced economic circumstances, health status and increased poverty for women. Conservative estimates suggest that women: make up two-thirds of the world's illiterate (UNICEF 2013); earn 10 per cent of the world's income (Global Poverty Project 2013); occupy only 18 per cent of seats in the world's parliaments (International Women's Democracy Centre 2008); constitute 70 per cent of those living in extreme poverty (Global Poverty Project 2013); and, with their children, represent 80 per cent of the world's refugee population (UNHCR 2005). These figures highlight the imperative to consider the unique circumstances that women face and how these factors can not only hinder their resilience and ability to cope with additional shocks and stresses in a post-crisis context but also impact their capacity to engage in environmental restoration projects as a first priority.

There is overwhelming evidence that women are more constrained during and after climate crises. UN WomenWatch notes that, in many contexts,

> women are more vulnerable to the effects of climate change than men [and] are especially vulnerable when they are highly dependent on local natural resources for their livelihood. Those charged with the responsibility

> to secure water, food and fuel for cooking and heating face the greatest challenges [and] when coupled with unequal access to resources and to decision-making processes, limited mobility places women in rural areas in a position where they are disproportionately affected by climate change. It is thus important to identify gender-sensitive strategies to respond to the environmental and humanitarian crises caused by climate change.
> (2011: 1)

Gender is one of the most significant indicators of vulnerability (Alston 2013; Lane and McNaught 2009; FAO 2007b; Neumayer and Pluemper 2007) and, in a crisis situation, this is exacerbated by a loss of control over natural resources, including water, the means of production, information, and decision-making; time poverty; a breakdown of educational and employment opportunities; increased exposure to unsafe conditions; and reduced capacity for local organizing (Dankelman 2010). Gender vulnerability to climate challenges is very evident in Bangladesh (Oxfam GB and Naripad 2011; Advancing Public Interest Trust (APIT) 2009; Cannon 2002).

In previous works (Alston 2013a, 2013b; Alston and Whittenbury 2012) and drawing on significant research conducted in the last decade (Dankelman 2010; Lane and McNaught 2009; Enarson 2012) I note that women are more vulnerable in post-disaster spaces because existing inequalities are overlaid with additional stressors.

As previously noted (Alston 2013a, 2013b) women are much more likely to die in disaster events (Neumayer and Pluemper 2007); to be exposed to violence during and after disasters; to be more constrained by their responsibilities for the aged and children (Enarson 2009, 2006; Dankelman and Jansen 2010; Alston *et al.* 2011); to be living in poverty, to have no ownership of land and resources, to have less control over production and income, less education and training, less access to institutional support and information, less freedom of association, and fewer positions on decision-making bodies (Dankelman 2010).

Research from across the world reveals the greater vulnerability of women during and after climate events, the greater likelihood that they will die, their increased morbidity, their enhanced exposure to violence and their lack of access to resources and decision-making bodies to influence and effect change (Alston 2013; Lane and McNaught 2009; FAO 2007b; Neumayer and Pluemper 2007). Women's vulnerability results from highly entrenched and normalized social practices and structural inequalities. For example, there are a range of customs that hamper the ability of women in Bangladesh to respond effectively to climate crises. Women's lack of land and resources limits their capacity. Women who participated in this research report not being able to go to shelters without the permission of their husbands; not receiving early warnings; there being no women's toilets in shelters; the threat of violence; a lack of privacy; not being consulted or provided with information; and being responsible for ongoing care work and income generation.

As slow-onset and catastrophic weather events become more frequent and more intense across the globe, gender differences in impacts, vulnerability, resilience and adaptive capacity are evident regardless of country, culture or event (Alston 2012, 2013a; Enarson 2009, 2012). Research emerging from climate-induced post-disaster sites reveals that the impacts for women are generally more severe than for men in terms of morbidity, mortality, ongoing livelihood restructuring, poverty and levels of support (see for example Alston *et al.* 2011; Lambrou and Nelson 2010; Enarson 2009; Harris 2010; Neefjes and Nelson 2010; Odigie-Emmanuel 2010; Vincent *et al.* 2010; Corral 2010; Tovar-Restrepo 2010). In the 1995 Kobe earthquake in Japan, for example, one-and-a-half times more women than men died; in the 1991 Bangladesh floods five times as many women died; and in the 2004 South Asia tsunami women were three to four times more likely to die (UNEP 2004, reproduced in Seager 2006). Understanding this greater vulnerability of women during and after climate events requires a sophisticated feminist analysis. Why are women more vulnerable? What are the factors that shape their responses? How does cultural context, level of poverty and family/household structure impact on their ability to respond to climate events with resilience?

Globally gender inequalities embedded in diverse cultural contexts lie at the heart of women's disadvantage. Examining these cultural and social practices in the context of climate change, enhances the possibility for gender sensitivity in climate and environmental actions; challenges discourses of *climate change* to incorporate social and political strategies; and reimagines post-disaster sites as critical spaces to address gender equality through actions and practices that not only acknowledge the rights of women and girls but also destabilize traditional customs that disempower them.

Nonetheless women's vulnerability is also embedded within a global context where social inequalities within and between nations are evident and growing. In the remainder of this chapter the factors that shape gender vulnerability are discussed. Connell's notion of a global gender order is introduced, as is the way in which women and girls are shaped in their cultural context by social practices that in post-disaster spaces may lead to significant disadvantage.

In summing up, Dankelman (2010: 59) also argues that women have less access to resources that are essential to disaster preparedness, mitigation and rehabilitation, and that their workloads increase not only because men are more likely to migrate to look for work but because of a lack of energy sources, clean water, safe sanitation and health impacts. Heavy workloads often result in girls dropping out of school to assist. Further, in previous research on the likely impacts for women, we conclude that there is/are:

- greater likelihood of deaths and injury during natural disasters;
- higher levels of physical and mental health issues;
- a greater role in caring for the sick and injured;
- a greater role in caring for sick children especially in relation to water-borne disease;

- a greater role in caring for the elderly, especially in relation to respiratory disease;
- a greater likelihood of violence and breakdown of societal protections following disasters and climate events;
- fewer employment opportunities;
- a greater likelihood that women will lose land rights;
- a higher rate of malnutrition as women tend to eat last;
- a greater burden of work collecting clean water and fuel for cooking;
- fewer roles in post-disaster reconstruction and decision-making;
- a greater loss of status due to declining participation in post-reconstruction, higher levels of violence, and a loss of basic freedoms;
- a lack of participation in household/community decision-making;
- relationship stress; and
- a loss of traditional women's knowledge.

(Alston and Vize 2010; Alston 2012)

Previous work with Bangladeshi women notes that they are compromised during disasters by a lack of access to information, food, safe water and sanitation, health care facilities, education, housing and ownership, land and inheritance, livelihood strategies, participation in decision-making, and protection, security and bodily integrity (Nasreen 2008). In particular, women in hill tribes and secluded coastal environments are more vulnerable due to their marginalized position and maltreatment from the mainstream community (Ahmed *et al.* 2008).

Policies and practices designed to address vulnerability and to aid adaptation are often gender-insensitive because women report not being adequately consulted (Denton 2002; Alston and Mason 2008). In the lead-up to the Copenhagen Climate Change forum, the UN Secretary-General Ban Ki-Moon called on world leaders to ensure that women are equally engaged in climate-change decision-making; that they be seen as agents of change and as custodians of knowledge essential to local natural resource management; and that the vulnerability of women in climate disaster situations be recognized and addressed (Ban Ki-Moon 2009).

Other world leaders support this view. For example, Amberoti Nikora, the Kiribati Minister for Internal and Social Affairs, during a speech to the 52nd session of the Commission for the Status of Women in 2008, noted the need for gender to be central to climate change negotiations because of the significant and complex outcomes in the Pacific region (Lane and McNaught 2009). Because there are so few women in decision-making relating to climate change, there is a significant lack of understanding relating to the gendered nuances of climate impacts. This can lead to an unintended reinforcement of gender inequalities through practices such as inequitable access to resources, aid, reconstruction work and governance practices.

Noteworthy is an almost total lack of attention to gender vulnerability in documents such as the first four IPCC Assessment Reports (2007), although there is more recognition in the fifth report. Arguably this gender blindness in global and national policy instruments and actions runs counter to a number of gender equality goals and actions sanctioned in other global forums.

A prevailing gender order

Understanding gendered factors that create uneven experiences is a critical factor in addressing the possibility for change. Gender inequalities are pervasive, global and reinforced by powerful institutional processes. In Bangladesh women are subject to a widely sanctioned and strictly enforced hegemonic masculinity. Men dominate land ownership, decision-making positions, production, leadership and commerce. Women are constrained through restrictions on movement, customs, laws and the dominant Islamic religion, and a widespread acceptance of violence against women, and practices such as dowry and early marriage.

Connell notes:

> The very gender inequalities in economic assets, political power, cultural authority and the means of coercion … means that men control most of the resources required to implement women's claims for justice. Men and boys are gatekeepers for gender equality.
>
> (2011: 7)

Connell further discusses a worldwide gender order that results in men being dominant in every society, reaping the rewards of women's subordination – a patriarchal dividend that he argues many are loathe to give up. Hegemonic masculinity is constructed, formed and reformed through social practices and transformed in the context of globalization. Men's resistances to gender equality measures range from passive to very active, with overt forms of violence, honour killings, acid-throwing and murder at the extreme end of a long continuum.

The gender order is a complex amalgam of various *gender regimes*, defined as the gender relations in various institutions such as the family and the state. In understanding the gender order, and women's and men's complicity with a system that is so obviously unfair, Bourdieu's theory (1970) of symbolic violence is useful. He describes this as a process whereby prescribed cultural symbols and meanings are imposed in such a way as to be seen by the oppressed as legitimate. He takes this further in his book *Masculine Domination* (Bourdieu 2001), where he argues that women and men tend to accept gender inequalities as natural and unchangeable, thereby justifying men's domination. In gender relations, *symbolic violence is the violence upon a social agent with his or her complicity* (Bourdieu 1992: 167). Lukes (2005) makes similar points in articulating his third dimension of power, a concept he describes as the notion that people who are subject to domination acquiesce to their subordination. Importantly he notes the role of institutions, such as the state, the family, schools, churches and the law, in perpetuating symbolic violence and gender inequalities. However, he points to second-wave feminism as responsible for making transparent the taken-for-granted in these accepted norms and customs and recognizing the potential of women's agency and resistances. Changing these social practices and cultural customs is critical to addressing inequalities (Hawthorne 2001).

Embodiment of social practices

During this research, I have become increasingly conscious of the embodiment of gendered social practices in the cultural context that is Bangladesh and the way that this marks and constrains women. Embodiment describes the way bodies are both objects and agents of customs and social practices (Connell and Messerschmidt 2005; Connell 2002). As Foucault argues, the body is culturally constructed as 'the field of inscription of socio-symbolic codes' (quoted in McLaren 2002: 92), producing what Foucault terms the 'docile body'. The docile female body is shaped through discourse, adornment, disciplinary practices and gender norms in ways that exert control (Bartky 1990; Bordo 1989; McLaren 2002). Shildrick (1997) adds that the female body in a male-dominated culture reveals gender inequalities and that the inscription of femininity onto female bodies stands in stark contrast to the disembodiment of masculine privilege. McLaren concludes that the body is central to our understanding of cultural norms and the representation of women as passive, weak and inferior. Haraway endorse this view when she argues:

> There is nothing about being 'female' that binds us as women. Gender, race or class consciousness is an achievement forced on us by the terrible historical experience of the contradictory social realities of patriarchy, colonialism, racism and capitalism.
>
> (1990: 197)

The embodiment of social practices is very evident in the Bangladeshi context. Married Bangladeshi women wear saris, a garment that restricts movement. They wear a nose ring bearing their husband's family symbol. They are expected to be subservient to male opinion and authority, to be escorted by a male relative when outside the home and to experience violence as unremarkable. During and after disasters this marking of the female body is problematic as clothing can restrict movement, their ability to reach shelters may be hindered by a lack of a male escort, and shelters may be unsafe. In post-disaster periods they may experience a lack of access to family planning support, unwanted pregnancies, a lack of hygiene products and unsafe sanitation.

The central tenet of this book is that climate change events make more evident the established gender order, symbolic violence, oppressive inequalities and embodiment of social practices. Lukes's point that these taken-for-granted power relations are more visible in abnormal times indicates that post-disaster sites can be potential sites to expose and challenge inequitable power relations. In the context of climate events and post-disaster reconstruction is it possible to challenge the taken-for-granted? Is it possible to ungender these sites, to reimagine the space, our relationship with the environment and the potent force of neoliberal economic policy? Is it possible to recognize and reward women's active agency, rights and needs?

Gender equality goals and actions

Despite the lack of gender sensitivity in climate-related policies and actions, major transnational gender equality frameworks have been developed in recent decades. Perhaps the most widely understood are the eight Millennium Development Goals adopted by the United Nations General Assembly in 2000. Four of these are particularly relevant to gender equality in the context of climate change – eradicating extreme poverty and hunger; promoting gender equality and empowering women; improving maternal health; and ensuring environmental sustainability. The target for the realization of these ambitious goals is 2015, following which there will be further actions on the development agenda and the crafting of Sustainable Development Goals. Transnational women's groups are active in lobbying for gender equality to be a central tenet of these goals. There is much work to do in an environment that is unstable and subject to extraordinary change. Keeping gender equality at the top of the development agenda and future development goals is critical.

Speaking at the Rio+20 Sustainability Summit in 2012, the then Head of UNWomen, Michelle Bachelet noted that:

> There is no longer any doubt that gender equality and the empowerment of women are fundamental to sustainable development. Solving our most pressing challenges requires our collective intelligence and wisdom. Given climate change, rising inequality and economic uncertainty, we need environmental protection, social equality and inclusive economic growth. And this requires women's full participation.

The Convention on the Elimination of Discrimination Against Women (CEDAW) is another gender equality measure that has received widespread global approval. Adopted in 1979 by the United Nations General Assembly it outlines an agenda to eliminate discrimination against women.

The third significant transnational initiative is the Commission for the Status of Women established in 1946. This body develops global policy to improve gender equality and the status of women. It holds an annual meeting to discuss issues of significance to women and to further develop policy to enhance the position of women. At the 2008 meeting a special panel session was convened to discuss the issue of gender and climate change; to note that climate change is not gender neutral; and to urge that a human rights framework be adopted in order to understand the full impact of climate events on women, their livelihoods, safety and health.

Fostering gender equality through climate change actions

Quite apart from the fact that addressing gender equality is a human rights issue, emerging research is also recognizing that gender sensitivity in post-disaster situations has positive social and economic benefits for women and men. It provides the basis for stronger economic growth and healthier communities. In developing countries women constitute 41 per cent of the global agricultural labour force (FAO 2007b) and undertake the majority of family-based care work (for example child care, aged care and food preparation). This care work extends beyond the home to include gathering fuel for cooking and collecting water. On average, rural women and girls spend almost an hour each day on these tasks and, in some communities, these activities may take up to four hours a day (FAO nd).

There is extensive international research suggesting that women exhibit greater awareness of, and responses to, environmental problems (see for example Goldsmith *et al.* 2012; McCright 2010; World Bank 2010) and that gender differences have been reported in pro-environmental attitudes, concern for the environment, knowledge levels regarding climate change and willingness to take action to improve environmental problems (Bord and O'Connor 1997; Semenza *et al.* 2011; Zelezny *et al.* 2000: 159–160).

International research also confirms that the engagement of women in post-disaster restoration significantly increases positive outcomes. The UN International Strategy for Disaster Reduction (2007: 9) notes that 'attention and resources directed to local, women's … traditional cultural knowledge and expertise produced more efficient, relevant and cost-effective projects'.

In summary, post-disaster sites provide a space for:

- building empowerment strategies for women into actions to protect the environment and enhance individual and community resilience;
- opportunities for women's participation in decision-making about the natural environment and for avoiding entrenching existing social inequalities; and
- improving women's protection from gender-based violence linked to natural resource use.

There is an urgent need for climate change frameworks and action plans, including IPCC reports and National Action Plans, to incorporate gender sensitivity, build empowerment strategies and foster gender equality goals. Without transparency and overt fostering of women's human rights, these strategies and actions risk not only cementing discriminatory practices and reinforcing inequalities but also overlooking the very significant environmental understanding and traditional knowledge that women can bring to global climate solutions. Actions should include consultations with women; involving women in decision-making; distributing resources including training and education to women; involving women in reconstruction, monitoring and early warning; providing services to women to assist with caring and re-establishment of livelihoods; fostering women's leadership; and resourcing local women's organisations.

Through these actions and strategies women can be empowered to gain control over their lives in all spheres including the family, the household, the local community, the marketplace, organizations, institutions and government.

Feminist theories of the environment

Drawing on feminist theories of the environment, I begin to articulate a new way of imagining equality in post-disaster sites. I use Ferree and Tripp's (2006: vii) definition of a feminist framework as being one that has the *broad goal of challenging and changing gender relations that subordinate women to men … and that differentially advantage some women and men relative to others.*

Eco-feminist theorizing

Many feminist researchers have grappled with the notion of women's relationship to the environment and the parallels between a dominance of nature and a dominance of women. Shiva in her and Mies's 1993 book *Ecofeminism* links the oppression of women with that of nature, arguing that the subsistence activities of women have been overlooked because of their non-monetary value in a capitalist system. She links patriarchy, capitalism and colonialism as oppressive factors in women's subordination and notes that ecofeminism is a feminism of ordinary women. Critiqued for essentializing women as close to nature and men as close to culture, Shiva notes:

> I have never seen eco-feminism as an issue of essentialism in biology. I do not say that women are genetically wired to be closer to nature but we are culturally wired because the patriarchal system pushed us out. It also told women that men had all the rights and the control but not the responsibility, whereas women had all the responsibilities but fewer rights.
>
> (Jahanbegloo and Shiva 2012)

In recent times ecofeminism has enjoyed a resurgence as women in various cultural contexts argue for greater rights to environmental assets for women.

Feminist political economy

By contrast with ecofeminism and with neoliberal ideology, political economy feminists such as Agrawal *et al.* (2009) argue a) that women's relationship to the environment is more complex than a simple equation between women and nature and b) that because neoliberal philosophy is harmful to humans we should prioritize human needs and well-being over capital accumulation. The increasing wealth–poverty gap is viewed as an indication that communities are failing to address human well-being. This perspective focuses on gender equality and argues that women's subordination runs counter to a healthy society (see for example Beneria 2003). Riley (2008) notes that power is central to inequalities and that male entitlement embedded in social and cultural systems limits women's autonomy and agency, creating systemic gender inequalities.

Agrawal *et al.*'s (2009) study of women in community forestry demonstrates that women face opposing pressures from their need for survival and their desire for conservation. Nonetheless she notes that women's presence makes a critical difference to all aspects of forest governance including effective participation, rule-making, violations, forest condition and the distribution of forest produce. Inclusive practices make sense.

The values underlying feminist political economy and articulated by Riley (2008: 5) provide a useful basis for a feminist framework of climate change activism. She notes these values as:

- human well-being as the foundational value and gender equality being central to human well-being;
- human rights, especially economic and social rights;
- women's personal autonomy within relationships of reciprocity;
- women's moral and political agency;
- recognition and valuation of women's work of social reproduction;
- embracing differences and eliminating discrimination – racial, ethnic, sexual preferences, class/caste, religious and national origin;
- ecological and environmental sustainability in the promotion of well-being and social reproduction;
- social cohesion and solidarity across families, communities, regions and nation states; and
- global common good.

A feminist construction of climate change and environmental disasters grounded in these values is useful in challenging the gender order and gender regimes within nation states.

Feminist political ecology

Like feminist political economy, feminist political ecology theorists reject an essentializing link between women and nature. Nonetheless, political ecology theorists take the view that gender is integral to the way the environment has been shaped and constructed. Enarson (2012: 37) notes that this view of the world 'focuses on how women have historically and in varying contexts drawn on the material circumstances of their lives, their social networks and their place-based environmental knowledge to respond to emerging crises, challenging both male dominance and environmental degradation in the process'.

Feminist political ecology adopts a more radical focus towards women's role and agency in ecological matters and sustainable development. Thus, as Enarson notes, gender inequalities, environmental degradation and disaster vulnerability are linked in such a way that ignoring gender risks perpetuating degradation and reducing adaptation. The everyday/everynight work of women noted by Smith (1992) embodied in the collection of food and fuel, and of growing and cooking food, ensures that they are attuned to their space and place. Seager (1996, quoted in Enarson 2012) notes that women are the first to notice if water becomes rancid, if children become unwell or if the family is in some danger to their well-being. Feminist political ecology centralizes women in the business of environmental issues, food and water security, adaptation and sustainability. This is evident in women's grassroots movements in various sites across the world where women protest against environmental problems linking environmental degradation to health issues, and organize to ensure that women's voices are represented in powerful positions and parliaments.

Feminist theories of the environment demonstrate the critical need for a gendered analysis of climate challenges to ensure the very survival of people under threat.

Collateral damage

Yet, as we examine these gendered impacts in Bangladesh, it is critical to note Bauman's (2011) new construction to our understanding of climate change. He co-opts the term *collateral damage* from its military context to define how climate policy solutions tacitly accept existing inequalities and inequitable distribution of resources. Using the case of Hurricane *Katrina* in the United States, he argues that those most affected were poor, black people who did not have the resources to flee the approaching catastrophe. Potently, and perhaps more controversially, he argues:

> A blood-curdling thought: did not Katrina help, even if inadvertently, the desperate efforts of the ailing disposal industry of failed humans, struggling to cope with the social consequences of the globalization of the production of a 'redundant population' on a crowded planet … the likelihood of becoming a 'collateral victim' … of any natural catastrophe is currently one of the most salient and striking dimensions of social inequality.
>
> (Bauman 2011: 7)

Bauman's claim is that inattention to vulnerability is a tacit acceptance that there will be people and indeed nations who are disposable. In the context of our work in Bangladesh, this is indeed a blood-curdling thought. Does scientific reductionism and inattention to the plight of critically affected nations like Bangladesh signal an acceptance of *collateral damage?* Further, as with *Katrina*, is there an acceptance within affected nations that there are people and groups who are dispensable? Given the extraordinary differences in mortality and morbidity between women and men, are poor, rural women the collateral damage of climate change?

Climate change – collateral damage in Bangladesh?

Although Bangladesh is a minor contributor to global warming with very low carbon emissions per capita (UNDP 2010), it is one of the most vulnerable (Khatun and Islam 2010; GoB 2005). The precariousness of its position results from its low-lying delta landscape and the consequent significant threat from inundation and flooding as sea levels rise and the Himalayan ice-melt continues. Two-thirds of the country is less than 5 metres above sea level and, by 2100, the coastline is expected to retreat by 10 kilometres, resulting in a loss of 18 per cent of its land base (World Bank 2000). These climate-related factors are exacerbated by cross-border issues including water rights disputes and by socio-economic conditions including high population density, poverty, and poor institutional capacity and infrastructure (Dasgupta, Huq, Khan, Ahmed, Mukherjee *et al.* 2010; Khatun and Islam 2010: 8; GoB 2005; Benson and Clay 2004).

Historically Bangladesh has been subject to floods, flash floods, droughts, salt-water intrusion, cyclones and storm surges. Climate change is exacerbating the scale, frequency and intensity of these events – droughts are considered to be more prolonged, temperatures more extreme, storm surges more intrusive, salinity intrusions are extending further inland, waterlogging is increasing, flooding is becoming more extreme and untimely, monsoonal rains are becoming more hazardous and extreme weather events are occurring more often and with greater intensity and duration. Over the past five decades there has been a significant loss of life, homes and livelihoods, and major disruption to infrastructure such as roads, transport, and services. Additionally there has been a marked decline in agricultural and fisheries productivity, a loss of coastal land, greater salinity intrusion into productive land, a loss of forestation and a marked degradation of ecosystems (IPCC 2007; Khatun and Islam 2010; Rashid *et al.* 2009; Ahmed *et al.* 2008; World Bank 2006; GoB 2005; World Bank 2000).

The links between these outcomes and climate change are made in The Intergovernmental Panel on Climate Change (IPCC) *Fourth* and *Fifth Assessment Reports* (IPCC 2007, 2014). These note that the number of extreme weather events will increase in South Asia, leading to major impacts on agricultural production and food and water security. These geophysical impacts will affect human health and well-being, lead to major population movements and have significant consequences for livelihoods and poverty levels. Sea level rises across the Bay of Bengal are predicted to impact millions of people, flooding their homes and leading to mass migrations. In summing up the impacts for Asia, the IPCC (2014) notes that temperatures will rise, water will become more scarce, food security will be threatened by declining production, coastal regions will be under increasing stress and climate changes will have increasing impacts on human health, poverty, livelihoods and security (IPCC 2014: WGII, AR5, Chapter 24).

As Pender notes:

> In terms of the impact of climate change few places in the world will experience the range of effects and the severity of changes that will occur in Bangladesh, which will include: average weather temperatures rising; more extreme hot and cold spells; rainfall being less when it is most needed for agriculture, yet more in the monsoon when it already causes floods; melting of glaciers in the source areas of Bangladesh's rivers altering the hydrological cycle; more powerful tornados and cyclones; and sea level rise displacing communities, turning freshwater saline and facilitating more powerful storm surges.
>
> (Pender 2008: 27)

Further, Bangladesh is one of the most disaster-prone areas in the world and, according to McSmith (2006), the cost to Bangladesh could be half of the US $58 billion ever received in foreign aid. The frequency of major events such as floods is increasing with each of eight catastrophic floods between 1974 and 2004 meeting the previous designation of a one in 20-year event (Pender 2008).

In a summary of major predicted changes in Bangladesh, Pender (2008) notes that:

- Temperatures will rise across South Asia with predictions being that they will be above the global average and as much as 3.3 degrees hotter (Christensen *et al.* 2007).
- Seasons are getting warmer in Bangladesh and will be about 1.2 degrees warmer by 2020 (Tanner *et al.* 2007).
- Monsoonal rainfall will be affected, becoming more intense, and the times when rain falls may vary.
- Rainfall will increase, making many areas wetter in winter and drier in summer (Tanner *et al.* 2007).
- As a result, flooding will increase across Bangladesh in the monsoonal season and reduce water in the rivers, affecting fishing in the dry seasons (Alam 2004).
- Areas affected by flooding will significantly increase (Ahmed 2006).
- River bank erosion is affecting 1,200 kilometres of riverbank, and approximately 8,700 hectares has been lost (Ahmed 2006).
- The number of major droughts is increasing, and the areas affected are growing, and this is seriously affecting food production (Selvaraju *et al.* 2006).
- Ground water availability is being affected (Roy 2008).
- The shrinking of the Himalayan glaciers will affect water in the rivers (Stern 2006).

- Bangladesh is the world's third most vulnerable country to sea level rises and this is likely to affect up to 30 million people (McGranahan *et al.* 2006).
- Salinity is encroaching into land and water supplies, increasingly exposing people to high salinity (NAPA 2005).
- Cyclones will become more intense and more frequent, with increased wind speed and rainstorms (Tanner *et al.* 2007).
- Hotter summers are creating violent tornadoes (Roach 2005).
- Extreme weather events will increase across South Asia, exacerbating mortality from extreme heat waves and cold snaps, and increasing disease impacts; during flooding, extensive areas of Dhaka experience major health crises as a result of raw sewage sludge pouring out of storm water drains (Alam and Rabbani 2007).

Additional complexities

India is proposing to build two hydro-electric dams in Meghalaya near the border of Bangladesh. These will have serious consequences for Bangladesh should the projects go ahead. Currently the governments are attempting to negotiate a water-sharing agreement and Bangladesh is requesting an impact assessment before any construction begins. Vidal (2013) suggests that there is a much more general and widespread water grab underway in Asia involving India, Nepal, Bhutan and Pakistan. Altogether 400 dams are proposed, with a further 100 being planned by China to divert water from Tibet (Vidal 2013). The Himalayas may become the most dammed region in the world, with significant impacts for countries such as Bangladesh. Water wars will reshape the impacts of climate events on Bangladesh, limiting its people's power and capacity to access clean water and to stabilize food security.

Climate change impacts

Nonetheless the impacts of climate challenges have been evident for some time. For example Bangladesh has the highest disaster mortality rate in the world (WEDO 2008). Between 1970 and 2005 there were 171 natural disasters (WEDO 2008) including floods, storm surges, river bank erosion, drought, arsenic contamination, earthquakes, fires and climate variability (Oxfam GB and Naripad 2011) and, on average, a severe cyclone strikes Bangladesh every three years (Dasgupta *et al.* 2010). The scale of these disasters is unprecedented.

In 1970, the devastating cyclone *Bhola* killed more than half a million Bangladeshis and another in 1991 resulted in the deaths of more than 130,000 people, 90 per cent of whom were women and children (Schmuck 2002). In the 1998 Bangladesh floods 1,000 lives were lost, 30 million people lost their homes, 700,000 hectares of land were destroyed and 11,000 kilometres of roads were damaged (ICAI 2011).

Approximately 5,000 people lost their lives in Cyclone *Sidr* in 2007. This cyclone affected 6.7 million people in 30 of Bangladesh's 64 districts (APIT

2009; Oxfam International 2008) and approximately 1.5 million homes were damaged or destroyed (Chughtai 2008), over 1 million livestock were killed and more than 2.5 million acres of crops were fully or partially damaged (Disaster Management Information Centre 2007). Numerous strategies and enhanced infrastructures have been implemented to reduce the number of disaster-related deaths. These include the provision of early warning systems, cyclone shelters and evacuation plans; the mobilization of volunteers; the elevation of houses; and the provision of food shelters (Magrath 2007; Oxfam International 2008; Benson and Clay 2004; Cannon 2002; World Bank 2000).

The consequences of more frequent weather events include:

- significant loss of life;
- major impacts on population health and well-being;
- decreased access to a clean water supply;
- food insecurity and widespread hunger;
- increased disease and health issues;
- mental health concerns;
- disrupted education;
- reduced agricultural employment and livelihoods;
- reduced income and consumption;
- reduced industrial output and labour productivity;
- income inequality and disrupted local and global economic activities;
- damaged infrastructure;
- pressure on facilities and support services;
- population displacement from coastal and island communities;
- internal and external migration potentially leading to 30 million climate refugees by 2050;
- social conflicts; and
- unplanned urbanization and slum growth in cities.

(IPCC 2007; Khatun and Islam 2010; Rashid *et al.* 2009; Renton 2009; World Bank 2006; GoB 2005; World Bank 2000)

Within Bangladesh it is acknowledged that those living in poverty, women, children, the elderly and the sick are the most vulnerable to climate challenges (GoB 2005, Khatun and Islam 2010; Pettengell 2010; Brouwer *et al.* 2007; Khandker 2007; IPCC 2007) and it is well established that vulnerability to climate change is directly related to gender and poverty (Pettengell 2010; Enarson 2012). The ultra-poor, many of whom are women, rely on climate-sensitive resources for their livelihoods, have limited assets in times of hardship, are least able to move from disaster-prone areas and are more likely to experience food insecurity (Renton 2009).

These impacts on the fragile nation of Bangladesh indicate the plight of less developed countries that have reduced capacity to adapt quickly and effectively. International negotiations on climate change continue to focus on economic impacts and productivity issues – the debate appears to have been hijacked by the developed world, particularly those countries keen to preserve and protect their economic advantage. Further, countries such as Australia, the US and the UK are winding back their commitments to greenhouse gas reductions. It is a chilling thought that we have reached a point where neoliberal economic outcomes have conquered social justice imperatives and that the goal of a fair and just world is rapidly receding. At present Bangladesh is largely forced to fight its own battles with some assistance from international non-governmental organizations (INGOs). With this in mind, I turn to the research.

About the research

The research that forms the basis of this book has been funded by an anonymous donation to the Oxfam-Monash partnership based in Australia. Oxfam Australia and Oxfam Great Britain (the lead Oxfam agent on Bangladesh) joined forces with Monash University to examine the impacts of climate change in this fragile country. Oxfam was particularly aware that there has been limited attention to slow-onset events such as droughts, river erosion and salt-water inundation. With support from the researchers from the Gender, Leadership and Social Sustainability (GLASS) research unit at Monash University, the project was developed and successfully funded by the partnership. The research was undertaken during 2011–2013 by a team led by Professor Alston from GLASS at Monash. The team worked in close contact with Oxfam partner organizations in Bangladesh, allowing significant logistical and access support.

The research focus was to assess *the gendered impacts of climate variability with a particular focus on slow-onset events.*

The research questions included:

- What are the gendered impacts of climate variability particularly in relation to slow-onset change and climate events?
- What are the experiences of women, men, girls and boys in relation to these?
- How do women, men, girls and boys make sense of their experiences?
- What strategies or adaptations do they adopt to address these gendered impacts?
- What policies, practices and mechanisms might be instituted by NGOs and government agencies at all levels to support women, to enable women to advocate on their own behalf and to ensure that plans to address climate variability, decision-making and resources address this issue?

Data was collected using both qualitative and quantitative methods. In March 2011 a scoping visit to Bangladesh was undertaken by the Monash team to meet with Oxfam Great Britain colleagues and partner organizations. During this visit, ten key informant interviews were conducted with representatives of large NGOs and government ministries located in Dhaka, and a preliminary visit was undertaken to the Barghuna field site.

Figure 1.1 Map of Bangladesh showing the areas where the research took place

Three regional areas were selected for subsequent in-depth study (see Figure 1.1). These included:

- Ghaibanda in the north with the partner organization, GUK Gana Unnayan Kendra. This is an area subject to river erosion and drought.
- Satkhira in the south-west with the partner Shushilan. This area is subject to salt water inundation and water-logging.

- Barghuna in the south with the partner Jago Nari. This is an area where there are many Char villages affected by river level rises. (Chars are low-lying, flood-prone villages either located on islands within rivers or adjacent to the rivers. They are particularly unstable and subject to flooding.)

In-depth fieldwork took place from October 2011 until April 2012 and included key informant and individual interviews at village level and focus discussion groups (see Table 1.1). As part of this process, four Masters students from the University of Dhaka joined the team to work with the Monash and Oxfam Great Britain researchers. These students undertook a

Table 1.1 Bangladesh – qualitative data summary

District	*Union*	*Village*	*Type*	*M/F/ Mixed*	*Total No.*	*<25*	*25–40*	*>40*
Gaibandha	Kamarjani	Kundorpara	FGD	Female	0			
				Male	3	1	1	1
			IN/KI	Female	2			2
				Male	3		1	2
	Kapashiya	Sripur	FGD	Female	0			
				Male	3	1	1	1
			IN/KI	Female	4	1	2	1
				Male	2			2
	Mollarchar	Shidhai	FGD	Female	1		1	
				Male	5	1	2	2
			IN/KI	Female	9	1	4	4
				Male	4		2	2
Satkira	Ishoripur	Srifolkathi	FGD	Female	1		?	
				Male	1		?	
			IN/KI	Female	1			1
				Male	5		2	3
	Koikhali	Sundorganj	FGD	Female	2		?	
				Malc	1		?	
				Mixed	1		?	
			IN/KI	Female	2		1	1
				Male	4			4
	Gabura	9 no Shara	FGD	Female	0			
				Male	2		2	
			IN/KI	Female	0			
				Male	4		3	1
	Munshigani			Female	5	1	2	2

Table 1.1 (continued)

District	*Union*	*Village*	*Type*	*M/F/ Mixed*	*Total No.*	*<25*	*25–40*	*>40*
Barguna	Baliatoli	Porirkhal	FGD	Female	1		1	
				Male	1		1	
			IN/KI	Female	0			
				Male	4		2	2
	Dholua	Noli	FGD	Female	1			
				Male	3			
			IN/KI	Female	3		3	
				Male	5	2		3
	Fuljhari	Tatulbaria	FGD	Female	2	1	1	
				Male	2		1	1
				Mixed	1			
			IN/KI	Female	3	1	2	
				Male	3		2	1
Dhaka			KI		10			
			Workshop		3			

Total key informants (KIs) 10; Dhaka

number of interviews and led focus group discussion groups in Bangladesh. In total, the team conducted 31 focus discussion groups and 63 interviews with individuals in nine villages across the three selected regions, and 10 interviews with key informants in Dhaka.

Two ideas-sharing workshops were conducted in Dhaka with stakeholders from governments, NGOs, women's groups and Oxfam partners in October 2011 at the beginning of the village-level data collection period and again in October 2012 to report back on findings. A final series of workshops were held in March 2014 to discuss and share the findings with key stakeholder groups.

During 2013 a large quantitative survey of 617 people across the three regions was undertaken in conjunction with Eminence, a Dhaka-based survey group. The survey took place in nine villages from nine unions in the three upazilas of Gaibandha, Satkhira and Barguna.

The villages where quantitative survey data was collected and the numbers of participants are listed in the Tables 1.4 and 1.5.

The sample of respondents was as shown in Table 1.3.

Final feedback was given in four sessions in March 2014 in Dhaka.

This is one of the most comprehensive studies of the effects of climate variability on local people undertaken in this part of the world. The scope and extent of the project and the involvement of so many partners has meant that the data is rich and comprehensive.[1]

Table 1.2 Survey area

District	*Upazila*	*Union*	*Village*
Gaibandha	Sadar	Kamarjani	Kunderpara
		Uttor Sreepur *(Initially planned for Kapasia)*	Sree Pur
		Mollar Char	Sidhai
Satkhira	Shyamnagar	Iswaripur	Srifalkathi
		Koikhali	Jadabpur *(Initially planned for Sundorgonj)*
		Gabura	9 No Sora
Barguna	Sadar	Baliatali	Porir Khal
		Badarkhali *(Initially planned for Fuljhuri)*	Tetulbaria
		Dholua	Noli

Table 1.3 Sample size by sample area

District	*Village*	*Women respondents anticipated*	*Men respondents anticipated*	*Total sample from each village*	*Total sample planned from district*	*Total sample collected from district*
Gaibandha	Kunderpaa	35	35	70	210	210
	Sree Pur	35	35	70		
	Sidhai	35	35	70		
Satkhira	Srifalkathi	35	35	70	210	197
	Jadabpur	35	35	70		
	9 No Sora	35	35	70		
Barguna	Porir Khal	35	35	70	210	210
	Noli	35	35	70		
	Tetulbaria	35	35	70		
Total sample collected: Women 298; Men 319; Total 617						

This research provides a major addition to our understanding of the gendered impacts of climate variability in very vulnerable areas of Bangladesh. It has incorporated almost equal numbers of women and men, and its location in three very different regions of Bangladesh that are subject to quite different threats makes this a rich source of data. This provides the basis for a unique global understanding of this very vexing issue.

Table 1.4 Sex and district breakdown

	Women	*Men*	*Total*
Gaibandha	106	104	210
Barguna	109	101	210
Satkhira	83	114	197
Total	*298 (48%)*	*319 (52%)*	*617*

Table 1.5 Sex and age breakdown

<15	*15–19*	*20–24*	*25–30*	*30–34*	*35–44*	*45–54*	*55–64*	*65+*	*Total*
W 1	17	37	52	46	81	40	20	4	**298**
M 0	7	23	39	47	74	55	46	28	**319**
T 1	*24 (4%)*	*60 (9.7%)*	*91 (15%)*	*93 (15%)*	*155 (25%)*	*95 (15%)*	*66 (10.6%)*	*32 (5%)*	*617*

Quantitative sample characteristics

Almost equal numbers of women and men completed the survey and ages ranged from 15 to over 65. See Table 1.4 detailing sex and district breakdown and Table 1.5 which indicates sex and age breakdown.

Most of the participants are married (546 out of 617) and only five women/girls are unmarried. Women tend to marry at a younger age and almost all by their late teens.

Occupation

Occupations vary in the three regions. However, the chief industry in Gaibandha is agriculture; in Satkhira it is agriculture and shrimp farming; and in Barguna agriculture and river industries such as fishing and transport. Men are predominantly engaged in agriculture, fishing, running small businesses, transport services and day-labouring. Women are more likely to note their occupation as household duties and day-labourers or tailors. Women are in fact more likely than men to be day-labourers, blacksmiths, poultry farmers and net knitters. Their day-labouring may include brick-breaking (a common sight in Bangladesh), soil-digging, and paddy-boiling. No women noted their occupation as doctor, religious leader or retired.

Main activities undertaken in 2012

In order to tease out work roles, participants were asked what were the main activities undertaken in 2012. Table 1.6 provides a summary of these activities.

Table 1.6 Main activities undertaken in 2012

Activity in 2012	*Women*	*Men*	*Total*
Farm own land	14	77	91
Raise livestock for sale	16	8	24
Raise livestock for food	27	11	38
Home gardening	7	5	12
Agricultural worker (local)	16	44	60
Agricultural worker (away)		20	20
Construction worker (local)	1	1	2
Construction worker (away)	2	4	6
Fishing	5	33	38
Shopkeeper	2	24	26
Driver	1	21	22
Ferry – boat	2	8	10
Service – government	10	8	18
Service – NGO	2	3	5
Service – private	4	6	10
Factory		2	2
Faith healer	4		4
Midwife	14		14
Childrearing	97	1	98
Cleaning house	203	13	216
Cooking food for family	230	19	249
Water collection	223	21	244
Fuel collection	216	23	239
Elected official	7	2	9
Study	19	8	27
Poultry farm			
Business	18	52	70
Carpenter	2	11	13
Soil-filling	16	30	46
Bagger	5	5	10
Day-labour	8	19	27
Shrimp-farming		22	22
Hut maker		17	17
Sewing	2	19	21

Limitations

There are always limitations in any research, and particularly a project of this size. These limitations include that this is a snapshot at a particular time, that there are other drivers of change that must be separated from climate impacts, that the data had to be collected to fit with religious festivals, weather events, seasonal activities and availability of team members. A particular constraint on two occasions was the need to work around a full-day strike called by religious fundamentalists to protest against the women's policy and other gains made in society. During these full-day strikes all transport ceases and violent confrontations are evident in the streets. It is dangerous to be in public. On both occasions when we were confronted by strike action, our team had to renegotiate interviews and survey appointments and remain off the streets.

Nonetheless, within the confines of these factors, we are confident that this is unique data that provides a deep understanding of the lives of people under constant threat of climate events.

Structure of the book

Chapter 2 outlines a brief history of the country that is now known as Bangladesh and gives an understanding of the culture and context for the study. Chapter 3 examines the policy frameworks that guide gender equality within the country, noting the extraordinary measures taken to advance women's status at the same time as culture and customary law act against their interests. In this Muslim country of 160 million people there is much to celebrate and there is much that reduces the status and power of women. Chapter 4 begins the presentation of the data gathered from the many participants who took part in the research. This chapter outlines what might be termed 'ordinary life' for the people in this very traditional, though rapidly changing, society. I examine customs and practices, livelihood strategies and living conditions for the people who took part in the research. Chapter 5 examines the perceptions of the participants relating to climate change. How do those most affected – rural people, dependent on agriculture and living in poverty in very isolated villages – perceive the changes? In Chapter 6 I relay the experiences of people affected by major disasters and outline the various stages from preparation, to managing the crisis, to the immediate aftermath and finally to long-term reconstruction. The impacts of climate change on the people of these vulnerable regions provide the substance for Chapter 7. How are food and water security, livelihoods, health and education affected by the major challenges of climate change? What adaptations are adopted by the participants to cope with climate challenges and what are their hopes and wishes for the future? Chapter 8 completes the study with an examination of the revolutionary potential for post-disaster sites to provide space for challenging inequitable gender relations. We note the need for a global women's movement to advance gender equality in actions and strategies designed to address this most challenging dilemma.

Summary

Climate change represents a significant crisis and demands concerted global action in respect not only of environmental and economic aspects but also the significant social implications. A global focus on climate science and technological solutions overshadows attention to people and communities critically affected by climate crises. This is particularly the case for nations such as Bangladesh that are subject to a number of challenging events, both catastrophic and slow-onset, all of which have impacts on food and water security.

Further, the particular vulnerability of women and girls in rural communities that are subject to major climate crises is noted. In countries such as Bangladesh, poor rural women have reduced capacity and resilience as a result of ongoing social practices, a lack of resources and influence, and limited involvement in decision-making. Nonetheless, emerging research indicates that women's local, traditional knowledge holds the key to positive adaptations. Including women in climate actions and adopting a gendered analysis in climate research is fundamental to global actions for future sustainability.

Note

1 The project was approved by the ethics committee at Monash University and was subject to several reports during the life of the project.

2 Bangladesh

Culture and Context

This chapter provides a contextual background to the research findings that follow. In part it describes the physical landscape and gives detailed insights into the socio-demographic profile of the Bangladeshi people. It demonstrates the vulnerability of the land and its people who are now coping with the additional, complex and multi-layered climate challenges.

Bangladesh is a fertile delta country located on the Bay of Bengal in South Asia (see Figure 2.1). It is surrounded by India on three sides, Nepal and Bhutan to the north and Myanmar to the south-east. The 580-kilometre coastline along the Bay of Bengal consists of marshland and jungle and is subject to sea level rises, storm surges, coastal erosion and salt water inundation. The country is largely flat except around the Chittagong Hill district in the south-east and the Sylhet region in the north-east and is only 820 kilometres from north to south and 600 kilometres from east to west.

Relatively small in area, it is nonetheless a highly productive landscape, with major river systems encompassing hundreds of rivers that flow from the north. These bear the massive water flow from the Himalayan mountains and India, down the majestic waterways across the vast delta plain into the Bay of Bengal to the south of the country. One of these rivers is the mighty Ganges that finds its way to the ocean at the base of this fragile country.

The massive movement of water from the Himalayas in the north and down through the country creates an unusual movement of sediment that shapes and reshapes riverbanks, creates new islands and land masses, and swallows others as it makes its way to the ocean. Hence there are no entirely accurate maps of the river systems. Some rivers are as wide as 20 kilometres during the monsoon season (Schmuck-Widmann 2001) and shrink during the dry. Estimates suggest that a billion tonnes of sediment are deposited along the more than 250 rivers each year (Centre for Environment and Geographic Information Services 2008).

The torrent of water leads to significant river bank erosion and, as the Himalayan ice-melt continues apace, this erosion is becoming more rapid and dangerous. Meanwhile the sediment forms new islands known as the Char islands that are quickly dotted with small villages where people settle for a time, growing crops and raising livestock. These islands are subject to flooding

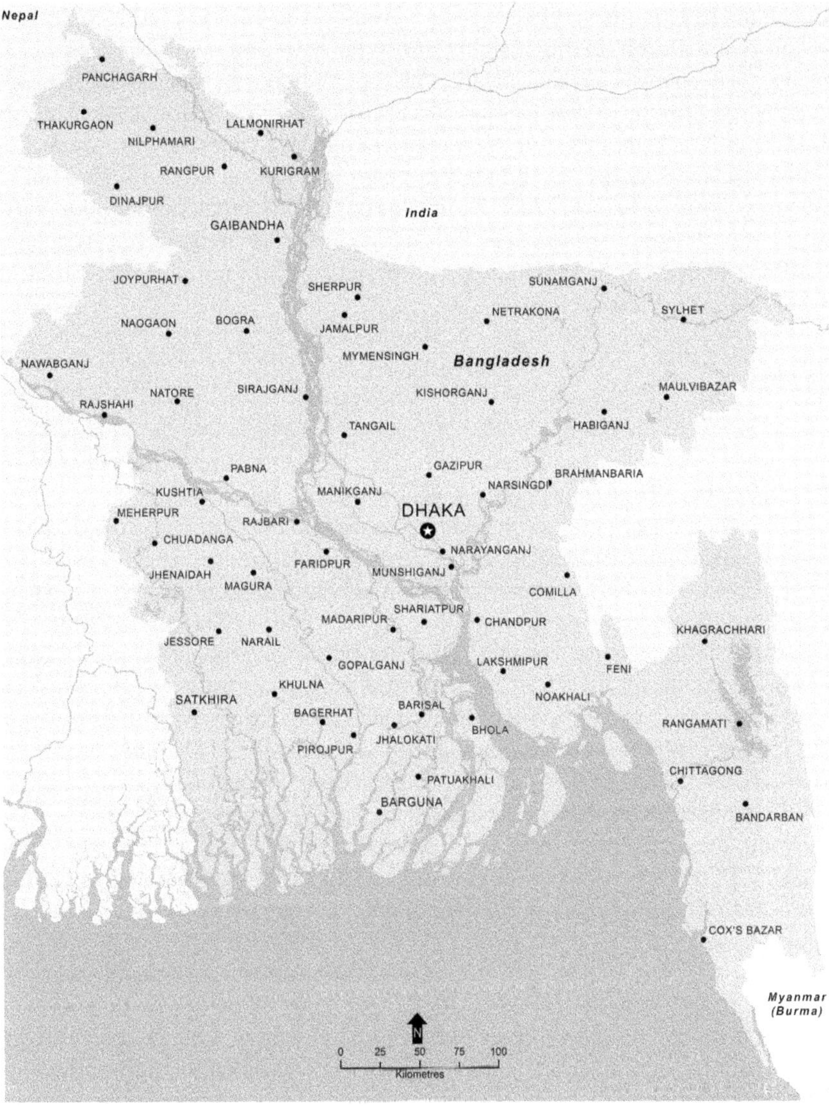

Figure 2.1 Bangladesh

during monsoons and as a result of other natural disasters. As the islands shift and change, so too do the people of the Char areas. Estimates suggest that 2.3 million people live on the Char islands in very isolated conditions and with limited service infrastructure. Many Chars are quite large – some up to 600 metres long (Schmuck-Widdmann 2001), but are mostly less than 20 years old, giving an indication of the transient nature of this population and the impermanency of their residences and livelihoods. Many Char people took part in this research, giving insights into their lives and circumstances.

The rivers themselves have sometimes changed course, as was the case with the mighty Jamuna River. It flowed east of the hilly region known as the Madhupur Tract until the 1700s when it changed its course to run to the west of this region. A major earthquake in India in 1762 or extensive flooding in 1787 may have caused this very significant change (Schmuck-Widmann 2001). Either way it demonstrates that the change and flux of the rivers is a long-standing phenomenon. Bangladesh is a country where nature reveals itself as both mighty and treacherous.

Industry

A large majority of Bangladeshis (72 per cent) are based in rural areas where agriculture and food production are the chief industries and occupations (UNDP 2010). Rice, wheat, jute and sugar cane are the main crops. Agriculture is the largest national industry, providing 22 per cent of gross domestic product (GDP) and employing 48 per cent of the labour force (BBS 2008: 10). Rural women contribute to economic activities within and outside the homestead, cultivating vegetables, rearing poultry, goats and cows, and taking over responsibility for farm management when husbands are absent (Paris *et al.* 2004).

Despite, or because of its clay soils and lack of forests, a visitor to these areas cannot help but notice the numerous brick factories with their towering smoke stacks dotted across the landscape manufacturing the ubiquitous red ochre bricks that form the building blocks of the rapid expansion of urban areas.

As industries and urban growth develop, unemployment, underemployment and land pressures have influenced out-migration of rural people to urban areas (NIPORT *et al.* 2009). Migration to the cities for work has become a significant option for rural family members, and female labour migration is a significant part of this movement (Naved *et al.* 2001).

Out-migration is evident in a decrease of the proportion of people who live in rural areas from 80 per cent to 72 per cent since 1980. Employment opportunities in the larger cities including the capital, Dhaka, are growing due to rapid modernization, major construction works and the hundreds of garment factories that service the West's insatiable desire for cheap clothes. The industries that feed off this growth, catering to the needs of a growing middle class, include the rickshaw drivers, roadside food stall holders, shopkeepers and domestic help. Rural family members find work in those burgeoning urban growth centres and send remittance income back to their families – income that is becoming critical to the sustainability of rural communities.

Since the 1980s, the Bangladeshi ready-made garment (RMG) industry has employed a significant number of rural women (Ahmed 2004; Kabeer and Mahmud 2004; Kibria 1995). As large transnational clothing companies take advantage of cheap labour in Bangladesh, these factories now provide 17 per cent of GDP, indicating the significance of this industry to national wealth and individual livelihood. The garment industry is a major employer of Bangladeshi women and provides much needed remittance money for families. The Bangladesh Garment Manufacturers and Exports Association (BGMEA) notes that, of the 3.1 million people currently employed in member factories, 2.38 million are women, most of whom are 'disadvantaged and economically poverty stricken' (BGMEA 2009). Women working in the factories are more likely to be single, divorced, widowed or deserted, or, if they are married, to have fewer children than non-garment workers (Kabeer and Mahmud 2004; Naved *et al.* 2001).

There is a dark side to this out-migration and urban development. Long hours and low wages, unsafe working conditions, human trafficking and a lack of safety, particularly for young women, are critical factors in this emerging modernization. This became very evident in April 2013 when a garment factory collapsed killing 1,127 workers who were trapped in the sub-standard building. This terrible tragedy has been described as 'the worst tragedy in the history of the global garment industry' (AP 2013) and followed a factory fire that killed 112 workers in 2012.

Women are recruited from villages as cheap labourers for these factories, work in insecure and unsafe conditions and experience a culture of fear and the threat of sexual harassment (Ahmed 2004; Siddiqi 2003). While women are empowered through increased earning capacity (Zaman 2001), they also experience institutional oppression as a result of unequal wages, sexual harassment and poor working conditions. There is a strong link between family honour and women's activities in public making it difficult for women to complain, mobilize or protest on their own behalf about their working conditions. Those who do may face social exclusion from their families in order to save face (Ahmed 2004; Salway *et al.* 2005).

Nonetheless the industry has been effective in providing new opportunities, but has not provided the working conditions and safety standards required for workplace security. Rapid industrial growth is not yet matched by safety regulations and, because labour laws are very weakly enforced in Bangladesh, labour rights are constantly violated (Berik and van der Muelen Rodgers 2008; CPD 2008a, 2008b; Absar 2003). Nonetheless Bangladeshi women view the garment industry as a way out of poverty and a vehicle to overcome oppression. As one NGO informant interviewed for this research notes:

> Of course, there are a lot more women now in education, the number of women who are in the garment sector, that's been a revolution in terms of women's mobility, and getting them out from the rural area, but, it's also an exploitative industry. So, the gains and the losses are difficult to measure. I know at one point we were very against having women come into this, but when we talked to women they said, 'But what choices do I have? It was about not having food and then coming [to Dhaka]. And the way you guys are talking about it, it's so theoretical because I was being exploited even when I was staying at home, or I was in the rural area.'

> So, our perception of right and wrong, black and white, is slightly different from if you are yourself in that situation. But [the garment industry] has helped women generally in terms of mobility. Because we have so many women garment workers today, there are more women on the street; women go out, women work more.
>
> (NGO Key informant, Dhaka)

There is no doubt that the garment industry provides much needed employment for rural women. What is required, however, is cooperation amongst stakeholders including the clothing manufacturers, government, union representatives and women's group leaders to ensure a safe industry and secure, sustaining working conditions.

Inequality

Poverty is a constant feature of Bangladeshi life – highly visible in urban and rural areas. Approximately 35 million Bangladeshi people live in extreme poverty (UNHRC 2005). A number of global indices give some insight into poverty levels and gender inequalities in Bangladesh. For example, Bangladesh is classified as a Lesser Developed Country on the Human Development Index and is ranked 129th of the 169 countries represented (UNDP 2010). Bangladesh also ranks near the bottom (93rd of 108) of developing countries on the Human Poverty Index (UNDP 2010) as a result of nearly half (49.64 per cent) the population living on less than US $1.25 per day. Nonetheless, because of the extremes of wealth, the average income in 2008 was estimated at $599, much of which results from remittance income sent by family members working elsewhere (BBS 2009).

On global gender equality measures, Bangladesh also ranks very low. The Gender Inequality Index places Bangladesh at 116th out of the 169 countries listed (UNDP 2010) and ranks the country 100th out of 128 on the 2007 Global Gender Gap Index developed by the World Economic Forum (Hausmann *et al.* 2007). Women are viewed as subordinate to men. An overt form of patriarchy pervades Bangladeshi society, dominating every aspect of life. Women are marginalized in every sphere including the law, recognition of human rights, economic participation, marriage and family customs, and land and resource control and they experience very high rates of gender-based violence. Purdah, the practice of shielding women from the view of other men, is widely practised, particularly in rural areas. Women's status, freedom of movement and autonomy is significantly restricted. Patriarchy in Bangladesh has been described as:

> A distribution of power and resources within families such that men maintain power and control of resources, and women are powerless and dependent on men.
>
> (Cain *et al.* 1979: 406)

History[1]

Despite being a relatively new country, the land now known as Bangladesh has a long history spanning at least 4,000 years. In its very early history it was a complex series of often feuding Indian city states. The earliest reference to this region in approximately 1000 BC referred to it as Banga. In more recent times it was part of India and known as East Bengal. Struggles for religious dominance between Hindus, Buddhists and Muslims have typified its long history. Buddhists ruled in its very early recorded historical period, but by the tenth century BC it was predominantly Hindu. The now dominant Muslim religion emerged from approximately the twelfth century, when Mohammed Bakhtiar Khalzhi from Turkistan captured Bengal. However, it wasn't until 1576, when the area became part of the Moghul empire, that the majority of the population of East Bengal converted to Islam. Until the sixteenth century Bengal was one of the most prosperous parts of the subcontinent and a site that was home to a flourishing and rich culture featuring art and music, trade and maritime pursuits (Bangla 2000, 2013; Highbeam Research 2005).

The East India Company set up a trade link in the late 1600s and came to dominate the area. The British Government took over from the East India Company in 1757 and controlled the area until 1947. Bengal became a fashionable and highly successful region in economic terms. However religion has played a large part in the area's ongoing history. The cooperation of Hindus with the British Raj and the lack of cooperation by Muslims resulted in the advancement of Hindus through British education systems and their widespread adoption of English. Muslims refused to be absorbed into the British system, resulting in widespread poverty amongst this group and a fractured social system.

World War Two saw the end of British colonialism and the creation of the state of India in 1947, a Hindu-dominated land. Religious conflict dominated the remainder of the territories. As a consequence the British government divided the area along religious lines – India became a Hindu state and the remainder – Punjab to the west and Bengal to the east of India – were named as one state – the state of Pakistan. That they lay nearly 1,000 miles apart was only part of the reason for the tensions that ensued. Their only binding relationship was the Muslim religion they shared (Mookherjee 2011). The dominance of the new state by the western Pakistanis – the Punjab – led to a determination by the newly formed Pakistani government that Urdu would be the official language, a decision that enraged the Bangla-speaking east.

Nationalist fervor was further fuelled when the Bengal-dominated Awami party received a majority in the 1971 election, giving Bengal the upper hand in government for the first time. However, the President of Pakistan refused to open the National Assembly and therefore blocked the new government from taking its rightful place. Riots broke out across Bengal and the people declared themselves the new independent state of Bangladesh. Pakistan sent its army into Bangladesh to quell the uprising. The sole purpose was to destroy Bangladesh, and they set about the task of genocide with ruthless precision.

First to be attacked was the University of Dhaka, where hundreds of students and teachers were killed and 7,000 citizens were massacred in a single night in March 1971. What resulted was one of the bloodiest battles of modern warfare. Within a week 30,000 people had been killed and 30 million were on the move, homeless as they attempted to escape the invading forces.

Bangladesh Liberation War

A critical factor in understanding Bangladesh and its people is to review this period of its recent history – a period that has now become known as the Liberation War. In 1971, for a period of nine months, the Pakistani army ruthlessly occupied the east, burning villages, systematically killing men and boys and raping women and girls across the territory. India, alarmed at the atrocities, provided support to Bangladesh, accepting up to 10 million refugees in the early days of the war. However, when Pakistan bombed India, Indian troops invaded Pakistan, hastening the end of the war and restricting the genocide. Once India entered the fray, this most brutal of wars was over in 11 days, but untold damage had been done to the people of the newly independent Bangladesh.

By the end of 1971, in one of the worst genocides of the post-war period, up to 3 million people, including dozens of the country's most prominent intellectuals, had been killed, up to 30 million people were homeless (Debnath 2009) and somewhere between 200,000 and 400,000 women and girls aged from seven to 75 had been raped, gang-raped and tortured by members of the Pakistani army and Razakars (local Bengali collaborators) (Mookherjee 2011: S74; Islam 2012). Many of these women were kept as sex slaves in Pakistani army camps for the duration of the fighting and many became pregnant. Islam (2012) estimates that approximately 170,000 women procured unlawful abortions after the war and up to 30,000 committed suicide.

Following the war, the government attempted to reduce the social ostracism being experienced by the raped women of Bangladesh, adopting the term *birangonas* (war heroines) to describe them (Islam 2012: 21; Mookherjee 2011: S79) and facilitating their entry back into society. They were given health care and training, and a *marry-off* programme was enacted to encourage Bangladeshi men to come forward and marry the women. However, as Islam (2012) notes, this was largely unsuccessful as they were portrayed as victims rather than as heroines. In her revelatory analysis of media commentary on the birangona during 1971–1972, Islam notes that they were described in newspaper coverage variously as *oppressed, shamed, ruined, soiled, tarnished, disgraced, annihilated, defiled and molested.*

Many of those who identified as birangona were ostracized by their families and communities. Others chose to remain silent, and Bina D'Costa (quoted in Islam 2012: 30) notes that women were compelled to silence about their experiences and 'a "negotiated survival" in order to avoid further persecution' (cited in Debnath 2009: 53). Many of these women are still alive today, keeping their stories to themselves.

A war crimes tribunal was established in the 1990s and continues to prosecute collaborators who took part in the war, and who murdered Bangladeshi men and boys and raped women and girls. In July 2013 for example, two collaborators were sentenced for crimes against humanity – one to death and one to 90 years in prison. In December 2013 Islamist leader Abdul Kader Mullah, a senior leader of the Jamaat-e-Islami party during the 1971 war, was hung for war crimes. A last-minute appeal forced a brief stay of execution and brought thousands of people onto the streets demanding his execution. The wounds inflicted by the Liberation War are deep and lasting.

In explaining the atrocities against the women of Bangladesh, Islam (2012) notes that women are the bearers of culture – they embody social practices evident in cultural identity. The use of rape against women is not only an attack on cultural identity, a genocidal act designed to break their spirit; it also identifies women's bodies as weapons of war. The establishment of the Bangladesh Liberation War Museum in the 1990s enabled the community to reflect and commemorate not so much the victory in war, but the involvement of ordinary people in freeing themselves. Its photographs and images of women and men killed or brutalized in the war are heartbreaking. Mookherjee (2011) feels that, apart from commemorating those who were killed or injured, one of the chief aims of the museum is also to ensure that radical Islamic ideals do not take hold.

This 'nation born out of human rights abuses' (Mookherjee 2011: S80) has adopted as its anthem a song called Amar Shonar Bangla (Golden Bengal) written by the Bengali, Hindu Nobel laureate poet Rabindranath Tagore.

Box 2.1: Amar Shonar Bangla

My beloved Bengal

My Bengal of Gold, I love you.

Forever your skies, your air set my heart in tune as if it were a flute.

In spring, O mother mine, the fragrance from your mango groves makes me wild with joy, ah, what a thrill! In autumn, O mother mine, in the full blossomed paddy fields I have seen spread all over sweet smiles.

Ah, what beauty, what shades, what an affection, and what tenderness! What a quilt have you spread at the feet of banyan trees and along the banks of rivers!

Oh mother mine, words from your lips are like nectar to my ears. Ah, what a thrill! If sadness, O mother mine, casts a gloom on your face, my eyes are filled with tears!

Governance

The new country of Bangladesh began, in theory, with a democratic style of government. Bangladesh's first Prime Minister in 1971 was Sheik Mujib, a prominent member of the Awami Party, who was later assassinated in 1975 in a military coup. The ravaged country experienced severe famine in 1973–1974, and the late 1970s were marked by political unrest, martial law and ongoing military coups.

In 1979 President Zia led a new democratic government that stabilized the country and effected good relations with the West. However, his assassination in 1981 led to renewed military governments and promises of a return to democratic rule – something that did not occur until 1991 (Khan *et al.* 1996). Political unrest between the major parties led to the boycott of elections and accusations of impropriety. New elections were held in 1996 and the Awami party again won a majority. The new Prime Minister was Sheik Hasina Wazed. She remained in this position from 1996–2001 and was returned again as Prime Minister in 2009. As opposition leader in the interim she was the target of an assassination attempt in 2004 (Riaz 2004) and was charged with murder and corruption in 2007 by the military-backed caretaker government and jailed for a year before charges were dropped (Hagerty 2008). She was eventually returned in a landslide victory in elections held in 2008 and remains a popular Prime Minister. Elections set for 2014 have again been disrupted by political unrest.

Theoretically Bangladesh has a parliamentary democracy with elections held every five years, and all citizens over the age of 18 are eligible to vote. There are 350 seats in the parliament with 50 of these assigned solely to women. The Prime Minister is appointed by the parliament and the President is appointed to a largely ceremonial position. The two dominant parties are the Awami League (a more left-leaning party) and the Bangladeshi National Party (BNP) (a more right-leaning party).

One national women's activist, in describing women's role in government, notes that women's seats are allotted by the party rather than their being democratically elected. This has resulted in political appointees rather than an empowerment base for women:

> Politics is very much a male show, but even slowly, slowly, women's leadership is developing … we are still fighting for direct election to the reserved seats in the National Assembly [to be elected positions]. We have already direct elections for the local level.
>
> (Key Informant, Dhaka)

There are several layers of government below the national government level. The next layer of government is the union level, somewhat equivalent to the Western notion of states or provinces – there are seven unions in Bangladesh: Barisal, Chittagong, Dhaka, Khulna, Rajshahi, Sylhet and Rangpur. These unions are then divided into districts or zilas of which there are 64. These zilas are divided further into sub-districts, or upazilas (of which there are 496), and each of these into smaller union parishads or councils consisting of several villages. There are 4,451 union parishads in Bangladesh. These are illustrated in Figure 2.2.

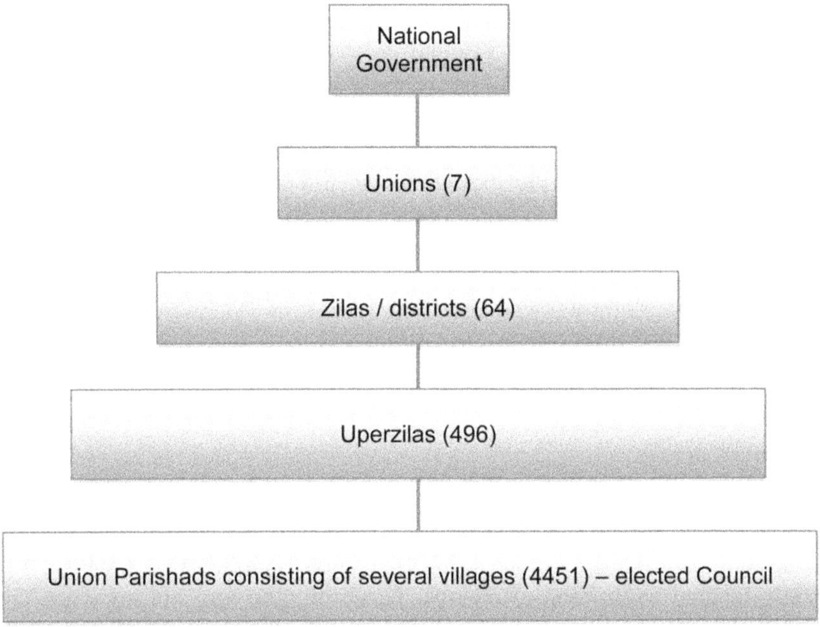

Figure 2.2 Layers of government

Union parishads are recognized as critical to local service delivery. The National Rural Development Policy describes these as the centrepiece of accountable government, of microcredit delivery, primary health care, disaster management and the control of communicable diseases (Aminuzzaman 2010). Union parishads have local elections to elect a chairperson and several members, and three out of every 12 of these must be female. However, there are no elected officials at the higher union or zila/district levels – government officials fill these layers of governance.

Demographic information

Population profile

There are approximately 160 million inhabitants in Bangladesh. This amounts to about 885 people occupying every square kilometre of its landmass (South Asian Association for Regional Cooperation 2008) making it the seventh most populated nation in the world despite its small size (Bangladesh Bureau of Statistics (BBS) 2011; United Nations Development Programme (UNDP) 2010). The largest cities in the country are the capital, Dhaka, with a population of approximately 10 million, Chittagong with approximately 3 million, Khulna with approximately 2 million and Rajshari with 1 million.

Bangladesh is one of the largest Muslim countries in the world with approximately 90 per cent of its people practising this religion, most of them Sunnis. The remainder of the population is Hindu (about 8 per cent), Buddhist and Christian. The country celebrates significant Muslim holy days including Eid-ul-Fitr, Eid-ul-Azha, Eid-e-Miladunnabi and Muharram.

The Bangladeshi population overall is concentrated in the younger age groups, confirming that life expectancy is lower than in many developed countries. One-third of Bangladeshis are under 14 years of age and only 5 per cent are 65 or over. The majority (56 per cent) are between 15 and 64. The ratio of males to females at birth is 1.04 : 1 (BBS 2009) and the male infant mortality rate is slightly higher than female. One in 15 children die before the age of 5 (a decline from one in 11 in 2004) and there is a significant statistical relationship between infant mortality and mother's education.

Indigenous people

Very limited information is available regarding Indigenous Bangladeshis. Yet there are more than 45 Indigenous ethnic communities comprising 3 million of the Bangladeshi population (Asian Indigenous People's Pact (AIPP) 2007), mainly located in the border regions. Those in rural and hill areas are considered the poorest people in Bangladesh, with 62 per cent of households below the absolute poverty line (Chittagong Hill Tracts Development Facility (CHTDF) 2009).

Indigenous Bangladeshis experience land dispossession, poor employment conditions, urban migration, discrimination, illiteracy, a lack of basic infrastructure, poor education and health access, and minimal participation in decision-making (CHTDF 2011; AIPP 2007; Roy 2000). Indigenous women are considered the most vulnerable section of Bangladeshi society and experience violations of human rights in many areas, including inheritance, legal and political rights, and a lack of decision-making powers. Further, Indigenous women are falsely portrayed as promiscuous, perhaps due to their more colourful dress (Guhathakurta, cited in AIPP 2007).

Indigenous women have experienced significant violence (including rape, kidnap and murder) by Bangladeshi peoples and military men through processes of displacement, and are particularly susceptible to sexual violence in their communities and in cities (CHTDF 2009; AIPP 2007). A study undertaken with sex workers in the city of Chittagong (Khan 2000) found that poverty, trickery, abuse, coercion and rape were identified as the main characteristics of entry into the sex industry. Seventy-eight percent of the participants in Khan's study had no formal schooling. AIPP notes that 'indigenous women are encountering severe economic hardship due to poverty, dislocation, loss of forestland and globalization' (2007: 21). Indigenous peoples appear disenfranchized and highly disadvantaged, despite the signing of the Chittagong Hill Tracts Accord in 1997, designed to protect their rights.

Refugees

In 2013 over 200,000 Rohingya refugees from Myanmar were in refugee camps along the border with Bangladesh being supported by NGOs. Although they have lived in Myanmar for over 200 years, they are recognized as Bangladeshis rather than citizens of Myanmar. Violent confrontations have forced them into neighbouring Bangladesh. However the government of Bangladesh does not recognize them either and has closed its borders to refugees.

Nonetheless, refugees along this border are common. In 1991–1992, for example, 300,000 refugees came to Bangladesh from the Rakhaine state in Myanmar, including many young single women who had been sexually abused on either side of the border (Masood and Wahra 1997). The Indigenous rights Chittagong Hill Tracts Commission (CHTDF 1991) investigated the refugee issues of the Chittagong Hill Tracts peoples in 1990–1991 and reported incidences of sexual attacks and rape (including gang rape) by the military and Bangladeshi settlers. Forced marriage was also being used as a strategy to integrate tribal women into Bangladeshi society.

Education

While one-third of Bangladeshi women and a quarter of all men have never been to school, approximately 61 per cent of men over 15 years and 52 per cent of women are literate (CIA 2013). There has been a marked change in school attendance over the last two decades and the educational status of women is improving. In 2000, there was gender parity in participation in primary and secondary education, and more girls are now enrolled than boys (Blunch and Das 2007; Hossain 2004). Boys tend to drop out of secondary school to provide labour in family enterprises, and girls for early marriage (Huda 2006; CPD 2009; see also UNICEF 2010).

Currently nine in every ten children aged 6 to 10 are in school and seven in every ten children aged 11 to 15 remain in school. From 16 onwards, urban young people are three times more likely than rural young people to remain in education and boys more than girls. Gender disparity continues at tertiary levels and there are 6 females for every 10 males participating at this level in Bangladesh (UNICEF and BBS 2010, cited in UNICEF 2010). Rurality and levels of poverty continue to determine educational access, and girls from poor rural families are more likely to leave school earlier (World Bank 2008). One of the critical strategies employed to improve the access of girls to education has been the Primary Education Stipends programme, begun in 2002, which benefited 5 million children in rural areas in 2005. This provides funding to the families of girls to enable them to stay in education. The female secondary school assistance project designed to improve the retention of girls beyond primary school has 2.3 million girls enrolled. These initiatives of the Bangladeshi government are to be applauded and are viewed as widely successful.

The reasons for educational gender inequity at higher levels of education are diverse but are generally related to early marriage and *purdah* (see Blunch and Das 2007; Sultana *et al.* 2009). There are significant risks associated with attending school, including public exposure and harassment – a practice euphemistically called 'Eve teasing'[2] in Bangladesh (Arends-Keunning and Amin 2000). Nonetheless, government policies and financial support to keep daughters at school has resulted in increasing numbers of girls and boys remaining in education at least until secondary level.

The importance of women's education cannot be discounted. Research suggests that improved education for women is associated with delayed marriage, increased contraceptive use, decreased family size, reduced infant mortality and lower general mortality (Nahar and Min 2008; Bates *et al.* 2007; Hurt *et al.* 2004). Bangladeshi women recognize that education decreases their vulnerability and increases their personal agency and bargaining power (Arends-Keunning and Amin 2000). Further, where possible, Bangladeshi people more generally believe that girls should be equally educated to or better educated than boys (Blunch and Das 2007).

Fertility and family planning

The 2007 Bangladesh demographic and health survey (BDHS) (NIPORT *et al.* 2009), reveals that fertility rates in Bangladesh have decreased markedly since 1975 when the average number of children per women was 6.3. In 2007 it was 2.7, indicating the success of the country's family planning programme and the widespread distribution of family planning information now universally available to women.

Family planning was first introduced to Bangladesh in the 1960s by health and social workers and adopted as government policy in 1965 in an attempt to reduce population. Population control has been identified as the top national priority by all governments since the 1970s, and the country has a target of achieving replacement size by 2060 (that is, nil population growth). As a result, 57 per cent of ever-married women use contraception – the Pill being the most common form. The most significant way in which Bangladeshi women receive contraceptive advice and products is via government-employed field health workers who until recently visited women in their homes. Nonetheless, the Bangladesh demographic and health survey revealed that 17 per cent of women had an unmet need for contraception, indicating that access is still an issue. Although abortion is illegal in Bangladesh (Singh *et al.* 1997), menstrual regulation is an accepted national family planning policy, which 6 per cent of the BDHS female respondents had used. Sterilization is also a popular method of birth control with the median age for this procedure being 26.

However, in recent times there has been a backlash resulting from men who have out-migrated to strict Islamic countries bringing back contradictory values. As one NGO informant reported:

> The success [of the government family planning campaign] didn't last very long because we started seeing the reverse. More and more, the size

> of the family is growing back again. My analysis – and a lot of other people are saying this, is that part of this is the fact that men started going as migrant workers to countries where they're much more orthodox in terms of use of contraceptive and family planning measures. So, they were coming back with these kind of messages … and insisting that God will feed as many children as he gives mouths – and so women were losing this right to practice choice.
>
> (NGO Key informant, Dhaka)

The change from visiting women in their home to the provision of satellite clinics has also had a detrimental impact on family planning:

> I also would say that our development partners messed it up. Because, there was this whole process where we used to have door-to-door family planning workers. In the seventies and eighties the success was attributed to that because they would come to your door, they would sit and talk to you and they would talk about contraceptives. Sometimes they would talk to the men and the men could be influenced and so on. These services were later, slightly removed. Because, the development partners deferred and all the others came up with the idea that now Bangladesh had achieved this much progress, it has matured, so we should have satellite clinics, we don't need to have this door-to-door service anymore – forgetting that women were not empowered to go to satellite clinics. But their argument was, yes, they would come for immunization, they would come with the children. Forgetting totally that women didn't have the cash to travel to these places. There was no transport, whatever, I mean there I could go on and on. Anyway the result of that has been that we have now gone back to a family size of five or six.
>
> Whereas the literate middle class it's one or two maximum. But where the 40 per cent of the population are below the poverty line, they're the ones. Some of the areas where I work, three generations have never seen a school. They all have children who are sick, six or seven children. They lose one to the water, they lose one to heat. You just can't be judgemental, but it's so difficult.
>
> (NGO Key informant, Dhaka)

Early marriage and childbirth

A further issue affecting women's lives is the practice of early marriage and the consequent age at first birthing. The legal age for women to marry is 18, yet most Bangladeshi women (78 per cent) are married by this age. The median age at which women are married is approximately 15. Despite family planning policies and increased female participation in education, Bangladesh has one of the highest rates of adolescent childbearing in the world (UNICEF 2010). In the 2007 BDHS, the median age at first birth was 18, and women reportedly had 23 per cent of their children before the age of 20 (NIPORT *et al.* 2009).

High rates of early childbearing are related to early marriage, low contraceptive use, age differences of more than 5 years between the wife and husband, and the social expectation that couples must have children soon after marriage to avoid a 'shameful' childless marriage (Nahar and Min 2008; Schuler *et al.* 2006; Kabir *et al.* 2001). Mothers-in-law also play a significant role in hastening early childbearing (Bates *et al.* 2007).

Adolescent childbearing is particularly unsafe in Bangladesh, with problematic access to services, increased risk of complications, higher morbidity and mortality. It also reduces access to education and work opportunities and thus results in lower status for women who marry at a young age (NIPORT *et al.* 2009: 56). Factors that can delay the age of first childbearing include the parents' (and mother-in-law's) education and frequent exposure to media (Nahar and Min 2008; Islam 1999).

Rural women are likely to have slightly higher fertility rates than urban women (an average of 2.8 children compared to 2.4). Nonetheless, women recognize the need to restrict family size because of poverty, food insecurity and the threat of malnourishment and disease. Most women (57 per cent) aged 15–49 express a strong desire to have no more children or several years between births (21 per cent). Furthermore, the median birth interval is 44 months, compared with 35 months in 1993 (NIPORT *et al.* 2009).

A large majority of women (85 per cent) have their babies at home. Several factors inhibit access to obstetric care, including limitations on travelling alone (NIPORT *et al.* 2003), wealth differences in accessing trained midwife services (Chowdhury *et al.* 2006) and medical costs (Koenig *et al.* 2007). Factors related to the use of facility-based maternal care include high socio-economic status, proximity to the facility, 10 years or more of education, higher age and being a member of a minority religious group (Anwar *et al.* 2004).

Those who use a health facility tend to be more educated and/or wealthy and are generally in urban areas. Only 21 per cent of Bangladeshi women have access to post-natal care. However, indicating that access to health care is improving, 82 per cent of infants are vaccinated by12 months and most children (98 per cent) are breastfed for on average 33 months.

Malnourishment

In Bangladesh food consumption and access to nutrition are critical issues affecting health and well-being. Twenty-six percent of Bangladeshis are malnourished (UNDP 2010). Malnourishment is very evident amongst children, but is also widespread across generations and in all communities. Approximately 43 per cent of children under 5 are defined as 'stunted' in their growth because of poor diet and a further 16 per cent as 'very stunted'. Seventeen percent are categorized as 'wasted' and 3 per cent as 'severely wasted', while 41 per cent are underweight and 12 per cent severely underweight (NIPORT *et al.* 2009). As one key informant describes:

> The nutrition of pregnant women is going to impact upon the next generation. Because the children in the womb and later on will be very much adversely affected. The so-called thousand days of opportunity will be lost. The gestation period and the first years of the life of children – that is the most critical. If you have deficiencies, deficiency of nutrition during that period, then this will never work well later in life.
>
> (NGO Key informant, Dhaka)

This severe malnourishment and impact on growth rate continues to adulthood. Women, in particular, have less-frequent consumption of nutritious food and smaller energy intake, and are therefore vulnerable to micronutrient deficiency (Sudo *et al.* 2004). Malnourishment and stunted growth has resulted in the average height of Bangladeshi women being 150cm. The 2007 BDHS identified that 30 per cent of women are chronically malnourished (NIPORT *et al.* 2009), while Rahman and Nasrin's (2009) study found that 43.7 per cent of participating rural mothers were malnourished. Women of reproductive age and children are the most vulnerable to malnutrition due to low dietary intakes, inequitable distribution of food within the household, improper food storage and preparation, dietary taboos, infectious diseases and poor care. Malnourishment is linked to place of residence, household wealth and particularly to level of education.

Social indicators that positively relate to better nutritional status of women include education level, the level of their husband's education and occupation, employment of women working for cash, intake of Vitamin A and iron supplementation, household assets index and spacing between births. Malnourishment of women is higher in rural areas.

Sanitation and health

Approximately 44 per cent of urban and 48 per cent of rural people lack access to improved sanitation, where human waste is separated from human contact, and 10 per cent of rural households have no toilet facilities at all. Consequently the risk of infectious diseases is high. The most common childhood illnesses are acute diarrhoea, respiratory infections and fever.

The health status of women is relatively poor in Bangladesh, despite a life expectancy of 69.7 years compared to 65.4 years for men (BBS 2009). Particular issues relating to women's health are nutrition, infectious diseases, and access to health care.

Patriarchy, enforced through *purdah*, limits women's access to health care as they are required to have a male escort. Consequently, women suffering illness seek health care significantly less frequently than men (Ahmed *et al.* 2000). Hadi (2009) found that 53 per cent of wives in the 2007 BDHS believed that their husband had the final say on their health care issues.

At the same time there is a widespread shortage of doctors in Bangladesh, with approximately 0.4 physicians per 1,000 people compared to countries like Australia where there are 3.9 per 1,000 people (World Bank 2013). Furthermore, the health system is poorly accessible to rural women who experience difficulties with health clinic personnel, including communication problems, conflicts over medication and uneven application of counselling skills (Schuler and Hossain 1998).

Women's employment

Women and girls engage in various types of paid and unpaid work, including household management, agricultural work, ready-made garment work, microcredit-funded small business development and the civil service. They are integral to the Bangladeshi economy, yet their contribution is largely unrecognized. Theoretically, men work outside and publicly, while women work inside the household undertaking domestic work, reproductive work, non-agricultural work and post-harvest activities (Hoque and Itohara 2008; Zaman 1995). 'Employment' is inconsistently defined within statistical reporting, but all reports show unequal rates of female employment compared to men.

Employment options for women and girls are varied but are very focused on agricultural production in rural areas, where 43 per cent of working women raise cattle, goats or chickens and 10 per cent work in home-based manufacturing. The growth of garment factory work is indicated by the 12 per cent of women now working in semi-skilled occupations. Furthermore, women who are more likely to be employed include those who are divorced, separated or widowed, with children, residing in poorer households, and aged 30–34 (NIPORT *et al.* 2009).

Nonetheless, women do not earn equal wages for equal work. One NGO is taking up the battle arguing that women should be paid a living wage. This ongoing campaign exposes the subordinate position of women in Bangladesh and customary laws which restrict women's equality and defuse equal rights campaigns:

> There is a discrepancy in terms of wages for women and men, not because of law, totally because of practice, custom. So, if a woman is working as a day labourer, she's working in the brickfield or she's on the roads, she's paid 80 or 90 taka, but the man gets 150 taka. Similarly, the wages of ready-made garment workers has also been an issue. [Women have] been asking for minimum wage and we've been campaigning not on minimum wage, but a living wage. Because the first round, we were really engaged with the minimum wage, and the minimum wage was fixed at 1,600. Then we realized that this was not [a living wage]. So, then there was a lot of lobbying and now it's been raised to 3,000, which again is inadequate, because it doesn't take into consideration education costs, health costs. So, there's a constant battle, conflicts around [women's right to a living wage].
>
> (NGO Key informant, Dhaka)

A high percentage (13 per cent) of children aged 7–14 are also involved in economic activity in Bangladesh (UNDP 2010). Boys are predominantly engaged in agricultural work, while girls are working in services, agriculture and manufacturing, and domestic service. The involvement of girls in household work is over ten times that of boys, and it significantly affects their school attendance rate and performance.

a) Household management

Women's unpaid household management role is categorized as 'housework' (this includes fetching water, collecting fuel, cooking for family and hired labourers, maintaining the home and animals, and caring for extended family members), 'motherwork' (childbearing and childcare, including caring for married daughters and grandchildren) and 'wifework' (satisfying the husband's sexual and emotional needs) (Chowdhury 2009; Paris *et al.* 2004). These roles are generally confined to the home. Amin's (1997) study in two rural villages found that women spent 60 per cent of their time in food preparation and other household work, and the remainder in animal care, fuel collection, child care and other tasks.

b) Agricultural work

Numerous studies demonstrate women's significant role in rural production and family subsistence, in work that is inadequately recognized, generally unpaid, invisible, and that results in limited decision-making influence (Mannan 2000; Rahman and Routray 1998; Amin 1997; Zaman 1996; Alam 1985). Men are responsible for making decisions regarding farm-related matters, managing the budget, making investments and dealing with public economic activities such as the marketplace (Hoque and Itohara 2008; Paris *et al.* 2004; Amin 1997; Kibria 1995; Ahmed 1992; Alam 1985; Feldman and McCarthy 1983).

Women have a significant role in post-harvest activities including the processing of crops, maintenance of gardens and livestock, and agricultural production, including food grain production and non-cereal crop production (Rahman and Routray 1998). During lean periods, women may engage in non-farm expenditure-saving activities such as making mats, stitching quilts, making fish nets, doing handicrafts, making clothes by hand, and making earthen ovens for parboiling rice (Paris *et al.* 2004).

There are indications that women's productive roles are changing in rural Bangladesh, due in part to landlessness, extreme poverty, weather variations and men's inability to provide an adequate income (Ahmed 2004; Kibria 1995). In 2000, nearly half of Bangladeshi rural families were considered 'functionally landless', owning less than 0.2 ha of land (insufficient for family income) (Hossain 2004). Other identified factors that influence women's workforce participation include wage rates, the development of village infrastructure such as electricity and the presence of NGOs (Bose *et al.* 2009).

c) Civil service

Despite a quota system dictating a minimum 15 per cent female employment for entry level positions (Jahan 1995, cited in Zafarulla 2000), there is a wide disparity between male and female representation in public administration jobs in Bangladesh. Bangladesh is ranked 90 on the Social Institutions and Gender Index, indicating poor inclusion of women in the civil service. Women comprise 10 per cent of legislators, senior officials and managers, and 22 per cent of professional and technical workers (Plan International 2009). In 2007–2008, women held 18 per cent of decision-making positions in the public sector, 12 per cent of first class government service positions, 8 per cent of second class government service positions, 9 per cent of positions in the bar council, and 16 per cent of positions in cooperative societies (BBS 2009).

d) Microcredit for small enterprises

Microcredit is a powerful movement in Bangladesh, providing loans to 20.6 million people in 2009, the majority being women (Mixmarket 2011). Microcredit, training and institutional participation have been found to positively influence women's participation in economic activities (Hoque and Itohara 2008), women's non-land assets (Chemin 2008) and the economic well-being of women and their families (Bernasek 2003).

However, microcredit appears fixed within a gendered framework. The Grameen Bank loans money to women for activities that are traditionally female, including raising poultry, raising cattle, weaving and food processing – activities which generally produce less income than men's activities (Bernasek 2003). Microcredit also increases women's workload by 9 hours per month, and women have higher repayment rates (Chemin 2008). While microcredit may reduce vulnerability, through asset-building, diversified income, consumption-smoothing and increasing women's asset control (Zaman 2000; Pitt and Khandker 1998), it may not reduce poverty (Develtere and Huybrechts 2005; Khandker 2005; Zaman 2000) and does not necessarily engage the ultra-poor (see Develtere and Huybrechts 2005).

Economic activity and empowerment

Nonetheless, there is significant evidence that access to microcredit increases women's empowerment (Kabeer 2011; Hoque and Itohara 2008; Chemin 2008; Pitt *et al.* 2003; Kabeer 2001; Zaman 2000; Nanda 1999; Hashemi *et al.* 1996). This occurs through the generation of income; greater economic status in the family and community; increased training and education; enhanced mothering skills; collective action against gender injustice; increased participation in familial and local decision-making; access to health education and intervention; increased mobility, networking and personal friendships outside the family compound; purchasing ability and autonomy; household-decision making ability; ownership of productive assets; legal and political awareness and knowledge of social issues; bargaining power and status in the household; access and control over household resources; spousal communication regarding family planning and parenting; and girls' school enrolment. Through microcredit women are also able to participate in communal activities that enable the exchange of ideas and a developing consciousness of gender relations (Kabeer 2011).

However researchers who take a more critical gender perspective argue that participation in microcredit results in increased economic security without increased empowerment (Rahman *et al.*, 2009; Mahmud 2003; Ahmed *et al.* 2001). Additionally, the Grameen Bank has not been successful in challenging entrenched gendered social institutions such as patrilocality (being required to live with the husband's family), divorce that favours men and dowry payment (Bernasek 2003).

Many working women do not control their income or loan, thus minimizing the potential autonomy and independence gained from economic activity (Chowdhury 2010; Hunt and Kasynathan 2002; Kabeer 1997; Kibria 1995). The 2007 BDHS (NIPORT *et al.* 2009) notes that only 31 per cent of currently married women with cash earnings decide how their earnings are used, compared with 56 per cent who decide jointly with their husbands and 12 per cent whose husbands decide. Urban, educated and wealthy women are more likely to decide independently how to spend their own money.

Women's work is considered supplementary to men's income and, in what appears to be self-protection against violence, women give their income to their husbands for household income management (Kibria 1995, 1997). Further, only a small proportion of women with microcredit are actually controlling their loan. Rather, they are purportedly passing on the finances to their husbands, sons or sons-in-law (Hunt and Kasynathan 2002). Chowdhury's study identified men's strategies to control their wives' income:

> they withdraw their financial support, they directly take all their wives' income, they save their own money while expecting their wives to meet family expenditure, they threaten their wives by saying that they will not allow them to continue their jobs, they buy land or flats in their own names by taking loans, repay the loan with their income and thus they compel their wives to meet family expenditure, they give false information about their loan to their wives and try to take all their wives' income, they attempt to use their wives' income to invest in businesses, and they appropriate their wives' income to maintain their parental family.
>
> (2010: 211–212)

Access to information

Women have very low access to publicly available information with nearly half (45 per cent) of all Bangladeshi women compared to 19 per cent of men having no access to mass media (newspaper, radio or television) on a regular basis. These statistics are particularly concerning given that access to information is critical for disaster warnings, aid distribution information and survival strategies.

Violence against women

Meanwhile violence against women is widely condoned in Bangladeshi society (NIPORT *et al.* 2009; Schuler and Islam 2008; Garcia-Moreno *et al.* 2005; Schuler *et al.* 1998). Schuler and Islam (2008) found that 84 per cent of women and 92 per cent of men approved domestic violence in at least one circumstance. The WHO multi-country study found that 53 per cent of urban women and 79 per cent of rural women in the Bangladesh study sites condoned intimate partner violence in at least one situation, and 63 per cent of urban women and 86 per cent of rural women who had experienced violence condoned the violence (Garcia-Moreno *et al.* 2005).

One-third of women and men surveyed for the 2007 BDHS (NIPORT *et al.* 2009) felt that there were some circumstances where wife-beating is acceptable. Nearly half (49 per cent) of ever-married women say they have experienced violence from their husband; 18 per cent of these women reported sexual violence; and 17 per cent being punched with a fist or something that could hurt them. A vast majority of men report that they abuse their wives, most commonly pushing, slapping or hitting. Rural women and women with lower levels of education are more likely to experience violence.

Only 25 per cent of the men who had reported being violent towards their wives believed hitting a woman was never justified. Reasons given for domestic violence include women failing in their family duties, disobeying elders, arguing with their husband, going out without telling their husband, visiting family or friends without permission, neglecting children, being economically dependent on men and refusing to have sex (Johnson and Das 2009; NIPORT *et al.* 2009; Hadi 2009; Schuler and Islam 2008).

Low levels of empowerment appear to be linked to women justifying domestic violence. The 2007 BDHS found that women who identified more reasons to justify domestic violence were less likely to participate in household decision-making, while women who identified no reasons to justify domestic violence were more likely to use contraception and access antenatal care, delivery service and post-natal services within two days of the birth (NIPORT *et al.* 2009).

Connected to high rates of justification of violence are low rates of help-seeking behaviour and reporting. Naved *et al.* (2006) found that only 34 per cent of women who had experienced physical abuse had told anyone about the abuse. The main reasons for not disclosing the abuse included:

> fear of jeopardizing family; honour; stigma that will tarnish the woman's own reputation; securing their child's future; fear of repercussion from the

> husband; hopelessness; expectation that things would change; threat of murder; and belief that violence is the husband's right.
>
> (2006: 8)

Community attitudes that appear to condone violence against women reflect patriarchal structures, very poor institutional and law enforcement to ensure women's safety and perceived helplessness of women in Bangladesh.

Marriage is considered the ultimate security for women, and therefore an abusive marriage is preferred because 'being unmarried poses great physical and social risks' (Chowdhury 2010: 204). Violence against women is so deeply engrained (Johnson and Das 2009) that it impedes women's participation and personal growth:

> In the face of actual or threatened violence, women cannot develop to their full human potential or participate in the economic, social, cultural, civil and political arenas on equal terms with men.
>
> (Azim 2001: 3)

Types of violence

Naripokkho's *Pilot Study on Violence against Women* (Azim 2001) outlines six categories of gender-based violence in Bangladesh: family violence; acid attacks; rape and sexual violence; custodial violence; murder and suicide; and community violence. The most common forms of violence, based on reports from newspapers and records of state institutions are torture for dowry, battering, threats, abduction, rape, trafficking, murder, abandonment, acid burn, burns from fire, 101 lashes or fatwa-related-community violence, and gouging eyes out of sockets. Most women who took part in this study (88 per cent) were acquainted with the perpetrator:

> In 2007 acts of violence against women reported to police included 3345 rapes, 147 acid attacks, 676 serious injuries and 9084 'others' – all of which were a decline from 2002 (BBS 2009). Between 1999 and 2010, there were reportedly 2433 incidents of acid attacks affecting 3144 people (the overwhelming majority were women), with only 476 convictions.
>
> (Acid Survivors Foundation 2011)

The WHO multi-country study shows that sexual abuse of women by non-partners was 7.6 per cent in urban areas and 0.5 per cent in rural areas. Physical abuse was 17.4 per cent in urban and 10.4 per cent in rural areas (Garcia-Moreno *et al.* 2005). There is also a rising incidence of public forms of violence against women, including rape, assault, acid-throwing and the punitive use of *fatwas* to enforce women's moral conduct (Kabeer 2011: 519) and increased evidence of trafficking of women for commercial sexual exploitation and forced labour (UNICEF 2009):

> My husband pressed me for dowry. But at that time my parents were unable to give that. We were just hand to mouth. They could only feed us but they were not able to give some thing to anybody. I was with my husband at his house; he used to beat me for money, for dowry. I sent a message to my father about the torture on me and asked for money. I told them that I could not stay there any more. How do I live like that! My parents tried to take me away from that home. They thought how long could I live like that?! How long will I be beaten by my husband and tortured? They waited for a long time and one day they took me back. After that they talked to me in detail. He asked me the reason why my husband is torturing me. I told that if I give him 70 thousand taka as dowry, I can continue to stay with him. If not I cannot live with my husband. My husband pushed me out from home and threatens me. My father was unable to pay that 70 thousand taka and I could not carry on the family any more. My situation was so bad!
>
> (Young woman, Barguna)

Many families refuse to take their daughters back in such situations.

Reporting of violence

Violence against women is significantly under-reported, and the actual rates are much higher than official statistics. Along with societal factors including social stigma, reporting may be hindered by structural challenges (three witnesses are required when a woman reports rape) and the fear of female sexual assault by police, which is allegedly frequent (Chowdhury 2009; Zaman 1999). Further, the fear of sexual harassment prevents many women from participating in the paid labour force or politics (Chowdhury 2009).

The reasons women give for not reporting violence include a perception that their lives will not improve (Schuler *et al.* 2008), social stigma (Hossain 2007) and fear of police assault (Zaman 1999). When women do react to violence, it is generally to take actions to 'avoid further violence by trying to conform to their husband's demands and expectations' (Schuler *et al.* 2008: 326). It is very difficult for women to leave their husbands, due to their social and economic dependence on men, the concern for children, the cultural shame and the impact of their decision on their unmarried female sisters (Bhuiya *et al.* 2003).

When women do disclose violence it is because the abuse included severe physical abuse, there has been a threat of murder and/or there is hardship caused to children. Only 2 per cent of respondents ever sought assistance from institutional sources (Naved *et al.* 2006: 1). Similarly, the 2007 BDHS reported that only 28 per cent of women who had recently (within the previous 12 months) experienced domestic violence told someone about the abuse and 47 per cent received assistance from someone (NIPORT *et al.* 2009). Despite these community attitudes towards violence, neighbours took initiatives to stop the abuse in 36 per cent of cases and arranged arbitration to mitigate the problems in 43 per cent of the cases.

Risk factors

The risk of violence is highest when:

- women are younger;
- women are younger than their husbands;
- both partners have lower levels of education;
- there is a history of abuse of the husband's mother by his father;
- there are problems with exchange of dowry, or unmet marriage demands;
- there is lack of an extended family structure;
- there is a lack of good spousal communication;
- women's participation in household decision-making is low;
- women are employed for cash as opposed to being unemployed or working without receiving cash;
- there are higher numbers of children;
- the husband is having extramarital sex, smoking bidi/cigarettes, chewing tobacco and using illicit drugs or alcohol;
- the couple are residing in a rural rather than urban location;
- the couple has low socio-economic status;
- there is economic dependency of the wife;
- the couple is Muslim rather than Hindu; and
- when a woman does not meet her husband's expectations in household work, or she disobeys or questions her husband, including regarding her husband's extra-marital affairs.

(Dalal *et al.* 2009; NIPORT *et al.* 2009; Bhuiya *et al.* 2003; Kahn *et al.* 2000; Aklimunnessa *et al.* 2007; Johnson and Das 2009; NIPORT *et al.* 2009; Hindin *et al.* 2008; Hossain 2007; Kanrunnahar 2007; Silverman *et al.* 2007; Huda 2006; Ahmed 2005; Naved and Persson 2005; Bates *et al.* 2004; Suran *et al.* 2004; Koenig *et al.* 2003; Hadi 2000; Schuler *et al.* 1998; Schuler *et al.* 1996)

Protective measures

Factors that may protect women from violence may include:

- having a bank account (Kahn *et al.* 2000);
- working in agriculture (Hindin *et al.* 2008);
- increased participation and appearance in public;
- group solidarity;
- education;

- skill-development regarding women's rights; and
- the increased economic dependence on women's income.

Some studies show a positive correlation between participation in credit groups and gender-based violence (Naved and Persson 2005; Suran *et al.* 2004), by as much as double (Bhuiya *et al.* 2003); some show a negative correlation (Hadi 2000), some a mixed relationship (Kabeer 2001; Schuler *et al.* 1998) and some no correlation at all (Bates *et al.* 2004; Ahmed 2005). Ahmed (2005), found that female members of the credit group BRAC experienced a greater level of domestic violence when they first joined BRAC, the violence reaching its peak when credit was introduced, and then substantially decreasing with skill-training over time.

Health and family violence

Family violence has numerous health consequences for women and children, including injuries (Bhuiya *et al.* 2003), the risk of sexually transmitted diseases (Decker *et al.* 2008), unwanted pregnancies and higher rates of infant deaths, pregnancy loss and abortion (see Hindin *et al.* 2008; Silverman *et al.* 2007; Khan *et al.* 2000). Additionally, women who experience violence by their current partners are less likely to seek antenatal care within the first three months of pregnancy and less likely to deliver at a health facility, and their children are more likely to be stunted (Hindin *et al.* 2008).

Summary

Bangladesh is a relatively new country with a very long history. It is a rich, fertile, densely populated river delta that is home to a largely rural population, most of whom live in poverty. Its struggles for independence and the emergence of a democratic government are testament to the will of the people to live in a free society. Yet there are very evident inequalities in this beautiful country, most evident between women and men, between urban and rural, between land owners and the landless, between the wealthy and poor, and the educated and non-educated. For many the daily struggle to survive is the constant theme of life. Yet very evident trends are emerging – the out-migration of rural people to the cities, of young women to the garment factories and of children into education, and lower birth rates. This is society on the edge of modernization and transformation. Yet, here, too, is a society where women are not free. From early marriage, to limited work options, to an acceptance of violence and of constant subordination to the men in their lives, women are subject to the whims of others. Can this change? Does our argument that climate disasters provide the space for changing gender roles hold true for a society such as this where rigid gender inequalities dominate every sphere of life? Before presenting the results of our research we move on to an analysis of Bangladesh women's policy and gender frameworks that may guide and assist change.

Notes

1 There are a number of significant works on the history of Bangladesh, notably van Schendel 2009 and Guhathakurta and van Schendel 2013.
2 Eve-teasing will be discussed in more detail in Chapter 4.

3 National and international policies and frameworks

Advancing gender equality

In this chapter the national and international policies and frameworks relating to women and gender equality are outlined to give an understanding of the commitments made within the country and in the transnational environment. Bangladesh is an active participant in world forums and has introduced strong measures in support of gender equality and women's empowerment.

Bangladeshi women live in a democratic society, albeit one with a troubled history. Nonetheless a moderate form of Islamic religion reinforces a discriminatory patriarchal system, despite their rights being recognized in the Constitution. This chapter provides a contextual background to the institutional frameworks that bind and restrict, and sometimes enhance, women's lives.

While democratic processes, the Constitution and national laws are based on the premise of equality of all citizens, the social practices and customs based on traditional Islamic law reduce women's access to freedom of choice and to control of their own lives. Women can vote, be elected to government, work in high-level positions and achieve their potential through education. They may also be married at a very young age to a much older stranger, be subject to dowry transactions, be a victim of violence, be constrained from owning property or the fruits of their labour, and be expected to undertake the majority of household tasks including collecting water and fuel, cooking, caring for children and raising livestock.

Particularly in rural areas, the lives of women in Bangladesh are circumscribed by these traditional customs, resulting in their social and economic position in the family and society being viewed as secondary. The social practices and customs dictate rules of residence and inheritance (Kibria 1995), entrench women's lifelong dependency on men, and place them under the authority of their father until they are married, their husband and family-in-law during marriage, and their sons or brothers when widowed (Sultana *et al.* 2009; Blunch and Das 2007; Ahmed and Bould 2004; Hashemi *et al.* 1996). Women's rights are rarely enforced. Nonetheless women's human rights are fostered through NGO activities, an active women's movement and Bangladesh's endorsement of transnational gender equality frameworks and a strong national Women's Policy. Nonetheless the prominence of NGOs and their influence on the national policy environment, and the tensions between equality mandates and customary practices, result in constraints on women's lives and women's activism.

The significance of the state

The complex layers of governance complicate the fight for gender equality in Bangladesh. Activists understand that the Bangladeshi state and system of government is critical to the facilitation of gender equality policies and their translation into practice. The state is a key institution that shapes and controls gender relations as it retains the power to provide leadership and support, and to create new language and frameworks that accept or reject human rights as a fundamental principle. Franzway *et al.* (1989: x) in their seminal work on the role of the state note that 'the state is an agent in sexual politics'. It is deeply implicated in shaping relations of power and is instrumental in forging particular views of women and men. For the most part it takes as its starting point a patriarchal view and develops and maintains gender constructions that lead to women's disadvantage. This is evident in countries across the world. State institutions, processes and actions reinforce a masculine norm, often treating women as a special case. Ferree (2012) notes that gender assumptions inherent in the state are a critical facet that shapes and reinforces gender inequalities. To achieve change in gender relations requires attention to the frameworks and actions of the state and to the way they construct 'women' and 'men' through legislative frameworks, institutional structures and actions.

The United States, the United Kingdom and Australia for instance tend to hold to principles of neoliberal individualism and equal rights, leaving citizen outcomes highly dependent on market forces and individual success. By contrast Ferree (2012) outlines the German model that is based on social justice and state responsibility for human well-being. Bangladesh's state structures reinforce a particular view of gender relations and yet demonstrate complex gender arrangements. The country's Prime Minister is female, despite the dominance of a moderate form of religious fundamentalism. Feminist activism is very evident in civil society as a voice of resistance, but so too are reactionary movements.

Women are still viewed as 'other' in many countries including Bangladesh, and the women's movement acts as a voice of resistance attempting, often unsuccessfully, to change the state's role in relation to gender inequalities. As Franzway (Maddison and Franzay 2006: 141) notes, 'the power of the status quo means that people will actually defend the illegitimate basis upon which their advantage is built'. During a field visit we witnessed this first hand when a violent, national strike followed the passing of laws giving women limited property rights. The pressure to maintain male privilege is overt, violent and dangerous.

The state is critical to shaping gender relations through its underlying philosophy, values, policies and practices and is, therefore, an important potential instigator of gender equality. Diverse positions in the way the state addresses gender relations in different nations can result in critical differences in women's experiences. The outcomes are highly dependent on the separation of religion and state, the recognition of citizenship rights, and in particular women's rights, and adherence (or not) to a neoliberal philosophy. In Bangladesh, women's rights are recognized in the Constitution, but so too is Islam as the state religion – thereby muddying the separation of powers. Nonetheless, Bangladesh supports a number of international frameworks that foster and support gender equality. Three are addressed here – the Millennium Development Goals, CEDAW and the Beijing Platform of Action.

Millennium Development Goals

At the UN Millennium Summit in 2000, 189 countries including Bangladesh signed up to the eight Millennium Goals (MDGs) aimed at reducing global poverty, hunger and disease by 2015. The goals are: eradication of extreme poverty and hunger; achieving universal primary education; promoting gender equality and empowering women; reducing child mortality; improving maternal health; combatting HIV/AIDS, malaria and other diseases; ensuring environmental sustainability; and forming a global partnership for development. There is no doubt that transnational and national attention to these goals has had a significant impact on these global indicators, although this has been variable across countries. In a widespread summary of achievements measured against the MDGs (ESCAP and UNDP 2012), Bangladesh is recognized as making gains in education access, reducing hunger, infant and maternal mortality rates. However it is described as slow on most other indicators.

Commenting on Bangladesh's achievements in relation to MDG 3 – promoting gender equality and empowering women – Bhuyan (2005) notes that the country has made some gains through programmes to improve food security and to provide access to training, day care, hostels for working women and advocacy to reduce violence. However, gender disparities are evident in tertiary education access, parliamentary representation, property rights, credit access, wage parity, employment, health indicators and violence.

Convention for the Elimination of All Forms of Discrimination against Women (CEDAW)

The Convention of the Elimination of All Forms of Discrimination against Women was introduced into the UN in 1979 and subsequently signed by Bangladesh on 6th November 1984, committing the country to gender equality and the empowerment of women. However, the government expressed reservations on four clauses that related to eliminating discrimination against women in relation to access to family benefits, equal rights in marriage and divorce, and rights to custody. These reservations were expressed on religious grounds. The groundswell of opposition to these reservations was immediate amongst women's groups, ultimately resulting in a softening of the national response.

Individuals and women's groups launched a CEDAW forum in Bangladesh in 1992 aimed at placing pressure on the national government to implement legislation in support of CEDAW. The Bangladesh National Women Lawyers' Association (BNWLA), Mahila Parishad and Nari Pokkho were amongst the women's groups that led this move. International Women's Rights Action Watch (IWRAW) assisted by providing workshops, training and a number of consciousness-raising forums (AWORC 1998). Strong lobbying within Bangladesh led to the eventual removal of the reservations imposed by the government on the contentious four clauses by 1997.

Beijing Platform for Action

The fourth UN World Conference on Women was held in 1995 in Beijing and resulted in a Platform for Action aimed at women's empowerment and gender equality. In recognition of the widespread nature of violence against women, the Beijing conference labelled this a critical issue in the Platform for Action and 'an obstacle to the achievement of the objectives of equality, development and peace' (UN 1995). The United Nations recognizes violence against women as a public policy and human rights concern (UN 1995).

Bangladesh has been an active participant in the international World Conferences on Women from the time of the first conference in Mexico in 1975. Following the fourth conference in Beijing in 1995, Bangladesh endorsed the Beijing Platform for Action. Ongoing lobbying by women's groups and NGOs within Bangladesh has kept a focus on gender equality and violence within the country, aided by these transnational declarations of equality and human rights for women and girls.

Human rights in Bangladesh

Within the country, human rights are enshrined in Section 3 of the Constitution of Bangladesh. This section recognizes the rights to equality before the law, equality between men and women and the prohibition of discrimination on the grounds of religion, race, caste, sex and place of birth. Eighteen fundamental human rights for all people are guaranteed (seventeen of which are directly from the Universal Declaration of Human Rights) and others are mentioned throughout the document. Additionally, Bangladesh is signatory to the International Covenant on Economic, Social and Cultural Rights (ratified 1998), the International Covenant on Civil and Political Rights (ratified 2000), and several others. Bangladesh also recognizes the right to safe drinking water and sanitation as human rights and has set national targets to achieve universal access.

In 2009 Bangladesh established a National Human Rights Commission mandated to address these fundamental rights. However, serious human rights abuses occur consistently throughout Bangladesh and have significant impacts on women (Human Rights Watch 2011; Asian Human Rights

Commission 2011). Women and girls are disproportionately represented amongst the poor and are prevented from participating fully in society because of cultural norms relating to marriage, violence, property, inheritance, dowry and early marriage and they are critically affected by limited access to safe drinking water and sanitation (UN Human Rights Council 2010).

Women and the law

Since the declaration of independence from Pakistan in 1971 and the formation of a parliamentary democracy, women have been recognized in the Constitution as equal. The Constitution of Bangladesh pledges the ideals of *nationalism, democracy, socialism and secularity*, and grants equal rights to men and women, and freedom from discrimination (s.28, People's Republic of Bangladesh 2011). Although Article 28 of the Constitution states, 'The State shall not discriminate against any citizen on grounds only of religion, race, caste, sex or place of birth' (Mitra *et al.* 2004: 211), women experience many forms of discrimination and inequality and have few protections, particularly against violence (NIPORT *et al.* 2009). While the original Constitution was a secular document, subsequent military regimes shifted the underlying principles to reflect Islamic law thereby undermining women's human and citizenship rights. While these changes were later declared by the High Court to be unconstitutional, the Constitution also recognizes Islam as the state religion and this conflicts with secular ideals of gender equality. Further, while the Bangladesh legal system is based on British Common Law, Islamic law is applied in personal and family matters and overrides women's equal constitutional rights regarding marriage, divorce, maintenance, guardianship of children and inheritance (Hossain 2003; Brandt and Kaplan 1995).

Islamic law

Chowdhury (2009) notes that the patriarchal (mis)interpretation and application of Islamic law in Bangladesh has resulted in the extreme subjugation of women. Briefly, Islamic laws prescribe that:

- wives are entitled to one-eighth of their husband's property, or one-quarter if childless, and daughters inherit half of the son's share;
- Muslim women may only marry Muslim men, however men may marry Muslim or non-Muslim women;
- men may practice polygamy with up to four wives, while women cannot practice polyandry;
- men must pay *dower* to the bride's family upon marriage (not to be confused with *dowry*, which is not considered within Islamic law, and prohibited under Bangladeshi law in theory but not in practice); and

- men may unilaterally divorce their wives for any reason, while women may only request divorce in very limited circumstances.

(Chowdhury 2010; Hadi 2009; Hossain 2003; Zaman 1999; Brandt and Kaplan 1995)

Common practices that reinforce female oppression include *purdah*, the wearing of the *burqa* (a traditional women's garment that covers the entire body and face except the eyes), and *fatwa* (a legal pronouncement of law, commonly understood as punishment for immoral behaviour).

Purdah is supposedly an enforcement of female modesty, and occurs through restricting women's movement outside the home, separating men and women, and imposing a dress standard to hide women's face and form (Parveen and Leonhäuser 2004; Amin 1997; Hashemi *et al.* 1996; Zaman 1995; Kibria 1995; Cain *et al.* 1979). The practice of *purdah* is strongest when a couple first marries, then gradually relaxes as the wife ages (Zaman 1995). Purity and honour are central to *purdah* and reflect patriarchal views concerning women's sexuality (Sultana *et al.* 2009).

When girls reach puberty they are forbidden to talk to or see unknown males or prospective grooms. *Purdah* is considered an 'instrument of repression' (Cain *et al.* 1979: 432), as it excludes women from economic and social life (Sultana *et al.* 2009; Amin 1997). It also restricts women's access to television, radio and newspapers, and other forms of information-sharing.

The *Burqa* is an extreme form of *purdah* and until recently was rarely seen in Bangladesh. Increasingly, female garment workers and middle-class university students are wearing *burqa*, as a radical expression of their participation in the minority counter-culture Islamism movement (such as Jam'at-i-Islami (Shehabuddin 2008)) and as a form of protection from sexual harassment, for personal and collective identity, and to maintain their middle-class status (Sultana *et al.* 2009; Rozario 2006). By contrast, *purdah* and the wearing of the *burqa* is diminishing amongst poorer populations, due to the need to work outside the home (Hossain *et al.* 2006).

Although prohibited by Bangladeshi law, *fatwa* is also practised throughout Bangladesh as a punishment for immoral behaviour. There are many examples of humiliating and torturous punishments of women for actions such as adultery, becoming pregnant while unmarried and, in some extreme cases, simply speaking to a man, voting, working with NGOs or accessing other institutionalized services (Shehabuddin 1999). Generally, the punishment for a married woman who commits adultery is harsher than that for a married man as the man can take his mistress as a second wife. Although *fatwa* is a process of gender oppression, its practice demonstrates distrust in the state and the ineffectiveness of the Bangladeshi legal system.

State laws

There are a number of laws designed to protect women, many with limited success. The state's role in the oppression of women and its perpetuation of oppression through the legal system is evident. The pre-independence 1961 Family Law Ordinance still applies and formalizes Islamic Personal Law. This requires dower to be paid promptly upon marriage, prohibits polygamy without consent from the Arbitration Council, extends divorce options to women, ratifies compulsory registration of marriage and requires arbitration for divorce. Forced marriage is prohibited in both Islamic and Bangladeshi law despite its widespread practice.

Additional laws specifically affecting women that give some insight into their lives include:

- the Registration of Muslim Marriages and Divorces Act 1974;
- the Dowry Prohibition Act 1980, making the exchange of dowry illegal;
- the Cruelty to Women Ordinance 1983, which prohibits kidnapping of women, trafficking, causing (or attempting to cause) death or causing grievous injuries to wives for dowry, with penalties including imprisonment from 14 years to life (Monsoor 1999);
- the Child Marriage Restraint Act (Amendment Ordinance 1984), raising the age of marriage to 18 for women and 21 for men;
- the Family Law Ordinance 1985, which addresses cases of marriage, divorce, custody and maintenance;
- the Repression against Women and Children (Special Enactment) Act XVII of 1995 to further address violence against women (Monsoor 1999);
- the Women and Children Repression Act (2000) and its 2003 amendment allowing capital punishment for rape, acid-throwing, dowry deaths, abduction, sexual harassment, trafficking, and prostitution (NIPORT *et al.* 2007);
- the Acid Crime Control Act 2002 (Ambrus *et al.* 2010; Hossain 2003; Rahman 1997);
- the Citizenship Amendment Act 2009 enabling women to pass on their citizenship to their children; and
- the Prevention of Domestic Violence Act 2009.

The Family Law Ordinance and other attempts to lessen the severity of Muslim Personal Law are inadequately enforced (Brandt and Kaplan 1995) and this is evidenced by the continued practice of polygamy; the almost universal payment of dowry to the groom's family; inadequate recognition and practice of dower; dowry-related violence against women; and child and arranged marriages (Gierbo and Imam 2006; Huda 2006). In many cases, women who experience domestic violence are not aware either of the laws against domestic violence or the assistance available from the public and private sectors. Others, aware of the inadequacy of legal protection and in order to ensure familial support and family honour, do not assert their legal rights.

For example, some women voluntarily waive their legal entitlement to half as much as their brother's share of the parental property to secure familial support if their marriage ends (Kabeer 2001).

Bangladesh women's policy

The Ministry for Women and Children's Affairs takes direct responsibility for women's policy. The National Council for Women's Development was established in 1995 to address national policy. This incorporates representatives of several Ministries and is chaired by the Prime Minister. As a result of these governance structures, Bangladesh's commitment to international covenants and charters, and the constitutional commitment to gender equality, the National Policy for the Advancement of Women (NPWA) was introduced to parliament in 1997 and adopted in 1998. This policy was developed in a spirit of cooperation between women's groups and government, and included women's equal right to inheritance. The policy commits the government to gender mainstreaming across all policies and ministries and promotes women's human rights, education and training, health and nutrition, political empowerment and the prevention of violence against women.

The conservative BNP-Jamaat party revised the policy in 2004. Words and phrases such as 'equal rights', the 'right to land' and 'inheritance' were replaced with phrases such as 'equal participation' and 'preference', thereby reducing the impact of the policy and removing women's right to inheritance. These changes were made to appease fundamentalist religious leaders who opposed freedoms for women. One women's group leader described these changes as a move from secularism to fundamentalism, a critical change for women's rights and freedoms.

The policy was strengthened again in 2008 and 2011 to give women greater rights and freedoms, and each time the revisions have been challenged by religious fundamentalists. Following the 2008 national elections, the new government declared Vision 2021, committing to mainstreaming gender issues in order to effect visible and substantial changes in women's lives.

The 2011 policy revisions further strengthened women's rights and committed to giving:

> women full control over their right to land, earned property, health, education, training, information, inheritance, credit, technology and opportunity to earn.
>
> (Yang 2011)

The national Women's Policy is based on the fundamental principles of human rights, gender equality, equal participation, safety and security, freedom from violence, the development of the girl child, equal representation, poverty alleviation, food security, health, housing, shelter, safety, security and freedom from trafficking, gender-sensitive disaster responses and the economic empowerment of women.

The 2011 National Policy for the Advancement of Women's goal for women's empowerment

This includes:

i promoting and protecting women's rights;
ii eradicating the persistent burden of poverty on women;
iii eliminating discrimination against women;
iv enhancing women's participation in mainstream economic activities;
v creating opportunities for education and marketable skills training to enable women to participate and be competitive in all economic activities;
vi incorporating women's needs and concerns in all sectoral plans and programmes;
vii promoting an enabling environment at the workplace: setting up day-care centres for the children of working mothers, career women hostels and safe accommodation for working women;
viii providing safe custody for women and children victims of trafficking and desertion, and creating an enabling environment for their integration into the mainstream of society;
ix ensuring women's empowerment in the field of politics and decision-making;
x taking action to acknowledge women's contribution in social and economic spheres;
xi ensuring women's social security against all vulnerability and risks in the state, society and family;
xii eliminating all forms of violation and exploitation against women;
xiii developing women's capacity through health and nutrition care;
xiv facilitating women's participation in all national and international bodies;
xv strengthening the existing institutional capacity for coordination and monitoring of women's advancement;
xvi taking action through advocacy and campaigns to depict positive images of women;
xvii taking special measures for skills development of women workers engaged in the export-oriented sectors;
xviii incorporating gender equality concerns in all trade-related negotiations and activities;
xix ensuring gender-sensitive growth with regional balance; and
xx protecting women from the adverse effects of environmental degradation and climate change.

The government has committed to resourcing all layers of government in order to address the NWPA. A National Women and Child Development Council has been established, as has a Parliamentary Committee and Gender Focal Points. Recognizing the importance of grassroots-level activities, there is a strong articulated commitment to organizing women at local levels, to assist with the coordination of their activities, linking them to NGOs and INGOs. Significant initiatives in the women's policy being implemented by the Ministry for Women and Children's Affairs include a commitment to:

- reducing female poverty through small and medium enterprise (SME) credit facilities to assist with marketing, skills development and employment opportunities;
- a Vulnerable Group Development Programme to provide food assistance and to develop training for ultra-poor women;
- a Widows Allowance, Elderly Persons Allowance and Lactating Mothers Allowance; and
- a Gender Responsive Community Policing Programme.

Despite the strong intent behind the policy, widespread violence and a national strike greeted the 2011 revisions. Protestors were angry that the amendments violated Muslim law, a position opposed by the women's movement. Women activists argued that the 2011 revisions were in keeping with the Liberation War and the Constitution. Nonetheless the protest was violent, widespread and spearheaded by Islamic fundamentalists.

The role of INGOs

International non-government organizations are playing a significant role in supporting human rights and gender equality in Bangladesh. Many organizations adopt these principles in overarching guidelines on policies and programmes. Aid-funding, distribution of services and empowerment strategies are making a significant difference to women's lives, although this is patchy and highly dependent on the loci of INGO partnerships and local partner organizations. INGOs have been responsible for enhancing girls' access to education, extending the reach of family planning services, improving access to health services and exposing violence against women, trafficking and child marriages. These organisations bring not only aid and programmes, but a firm ideological commitment to gender equality supported by extensive resources, personnel, partnerships and outreach.

The Association of Development Agencies in Bangladesh (ADAB) was founded in Bangladesh in 1974 as a loose association of NGOs working in the country. It is now a strong and effective coordinating organization for in-country NGOs ensuring maximum cooperation and benefit and is strongly committed to gender equality. Khushi Kabir, coordinator of Nijer Kori, a national NGO, and a Nobel Peace prize recipient in 2005 for her work with landless Bangladeshis, is a former head of ADAB. The NGOs have made a significant commitment to consciousness-raising at village level and amongst poor women.

National- to local-level representation

Nonetheless, women are significantly under-represented at all levels of governance. At national level, despite both major parties being led by women, overall representation is approximately 2–3 per cent. While 50 seats are reserved for women, these are allocated by the elected parties rather than contested, and thus this has not addressed gender parity in elected seats, in political party administration or among office bearers (Jahan and Amundsen 2012). Women activists are demanding direct elections for women's reserve seats and a quota on parliamentary nominations.

Meanwhile the most important layer of government for local service delivery is the village-level union parishads. These organizations have a number of factors working against strong governance practices at local level. Chief amongst these is the exclusion of women from decision-making processes (Aminuzzaman 2010). Others include limited understanding of the role of the parishads, a low awareness by members of their official roles, the significant power of the Chair, limited revenue base, a lack of accountability, highly centralized programme design, a poor relationship between the administration of local elected representatives and increased power of MPs over the local elected representatives. Critical services such as health and education, family planning and nutrition come from central government with no involvement or advice from the union parishads (Aminuzzaman 2010). This results in significant dissatisfaction amongst the villagers in relation to the work of the parishads.

Nonetheless this layer of governance has the capacity to mobilize people and resources in a crisis, to be more responsive to local needs and to become a hub of development activity and service delivery (Aminuzzaman 2010). In particular, this level of governance could provide significant empowerment for local women through a stronger and more effective involvement of women at this level, transparent and accountable service delivery and empowerment strategies at local levels, consultation and advice from local women and men on ways to improve their lives and on the distribution of aid, services and development, and more overt and equitable representation of women.

The lack of representation of women in decision-making at all levels has a significant impact on the issues that are addressed, the way resources are distributed and the influence of various people in the community. Without this representation at local to national levels, women have limited ways of influencing policies and practices that disadvantage them. Further, the weakness of governance structures from national to parishad levels results in widespread opportunities for corruption.

Given the international covenants, the constitutional recognition of equality, state laws and policies, the role of NGOs and women's low levels of representation, how are women's lives circumscribed or advanced by the many obligations to which Bangladesh has committed? Using the examples of marriage, dowry, divorce and family structure, it is evident that gender equality and women's rights continue to be very constrained.

Marriage

Marriage and wifehood are viewed as natural vocations for women in Bangladesh. The universality of marriage (Amin 1998; Kabir *et al.* 2001; Chowdhury 2009) is expressed as a cultural norm and is rarely challenged. Marriage is considered crucial for women's status, socio-economic security and personal protection (see Kabeer 2011; Rozario 2006; Bates *et al.* 2004; Zaman 1999; Kabeer 1997).

Parents, guardians or professional matchmakers arrange marriages, and women do not generally participate in negotiations (Huda 2006; Schuler *et al.* 2006; Gierbo and Imam 2006; Bhuiya *et al.* 2005; Zaman 1995). Lineage, family background and reputation are considered fundamental for women seeking partners. There is some suggestion that men prefer brides with fair skin, beauty and docility, who are young and chaste, and who are skilled, obedient and able to adapt to the husband's household. More recently their capacity to earn income and their education have become important (Gierbo and Imam 2006; Huda 2006).

By contrast, a high-quality husband is viewed as one who is a good provider, and this is reflected in the cultural norm of women 'marrying up' in Bangladesh – wives usually have a lower social status, caste, employment status and educational level than their spouses (Blunch and Das 2007; Gierbo and Imam 2006; Arends-Keunning and Amin 2001). Men are often up to 10 years older than their spouses, a trend that often results in early widowhood (Dalal *et al.* 2009; Nahar and Min 2008; Schuler *et al.* 2006; Alma 1985; Cain *et al.* 1979).

Because nearly 80 per cent of girls are married by the time they are 18, and the median marriage age is 15 (NIPORT *et al.* 2009), early marriage is linked to low education levels for mothers and daughters and poverty (NIPORT *et al.* 2009; Bates *et al.* 2007; Kabir *et al.* 2001). Poor and ultra-poor families are more likely to marry their daughters at a very young age (as young as 9) to ease the family's financial burden (Schuler *et al.* 2006).

Parents give many reasons for marrying their daughters at a young age, including: to protect girls from rape and sexual harassment; the cheaper dowry expectations; the desirability to 'shape' a young girl before they become too independent; young men's sexual desires; avoiding the risk of girls participating in premarital sex and potential gossip which influences marriageability and dowry; myths of young women's greater sexual and procreative power; young women being considered more physically attractive, adaptable and compliant; avoiding the societal dishonour of having an unmarried daughter; and women older than 25 being perceived as less fertile (see Chowdhury 2010; Rozario 2006; Schuler *et al.* 2006; Suran *et al.* 2004; Arends-Keunning and Amin 2001). Essentially these reasons relate to 'honour' and financial considerations.

The negative health and economic consequences of early marriage and childbearing include generalized poor health, maternal and child mortality and morbidity, compromised ability to work, impoverishment, lack of empowerment, lack of familial harmony, domestic violence and marital breakup (UNICEF 2010; Schuler *et al.* 2006). There is a growing community awareness of these negative consequences, a changed societal perception regarding women's education and employment, and stronger support for delayed marriage (Schuler *et al.* 2006). Notably, demonstrating the link between early marriage and economic factors, the age at marriage is greater for women who are garment workers (Naved *et al.* 2001).

Dowry

Marriage in Bangladesh is often described as a 'transaction' (see Suran *et al.* 2004; Lindenbaum 1981), due, in part, to dowry: the practice of the bride's family giving large sums of money, jewelry and other goods to the groom's family – thereby paying for the daughter's 'security, happiness and a good and timely marriage' (Chowdhury 2010: 203) and security against divorce (Gierbo and Imam 2006).

Dowry is a regional cultural practice rather than a religious one. It appears to have begun in the late 1960s. Prior to this the *pon* (or bride price) was practiced (Alma 1985; Lindenbaum 1981) and the groom's family gave money to the bride's. Despite being outlawed in 1980, the practice of dowry is very common, and increasing as male employment becomes more unstable (Blunch and Das 2007; Huda 2006; Suran *et al.* 2004; Bates *et al.* 2004). Dowry is negotiated through a marriage broker, and the size and payment is affected by age and wealth of the bride and groom, status, the bride's physical characteristics, the perceived comparative quality of the bride and groom, the bride's education and employment, and family alliances (see Chowdhury 2010; Gierbo and Imam 2006; Huda 2006; Suran *et al.* 2004).

There is a financial incentive for early marriage as a one-year higher age at marriage can increase the value of a dowry by nearly 50 percent (Do *et al.* 2006). The price of dowry is rising (Ambrus *et al.* 2010; Huda 2006; Suran *et al.* 2004) and is correlated to family planning programmes and subsidized contraception (Arunachalam and Naidu 2006). Dowry has resulted in increased greed amongst men, who may leave their wife to seek better dowry from another wife (Alma 1985). Nonetheless, self-arranged or love marriages are increasingly common and generally do not involve dowry (Gierbo and Imam 2006).

The practice of dowry specifically affects the poor, and particularly families with multiple daughters. Many families must sell land or take loans and, if dowry payments cannot be met, daughters are often beaten and/or returned to their family home, potentially remaining unmarried and a cause of shame and dishonour for their families (Simmons 1996). A further risk of not paying dowry is that a daughter may remain unmarried (Huda 2006). Thus, while many

Bangladeshis condemn the practice (Schuler *et al.* 2006; Bates *et al.* 2004; Suran *et al.* 2004), most conform (Kamrunnahar 2007). Consequently, 'girls have become a rope around their parents' necks' (respondent, cited in Simmons 1996: 258), and are 'liabilities for their family' (Chowdhury 2010: 198).

The practice of dowry increases women's vulnerability. Dowry-related domestic violence is frequent and increasing (Chowdhury 2010; Hossain 2007; Huda 2006; Suran *et al.* 2004; Bates *et al.* 2004) and includes dowry-related acid attacks, where young women of families who are unable to pay dowry are attacked with sulphuric acid, severely disfiguring their faces and bodies (Acid Survivors Foundation 2011; Kabeer 2011). Many suicides of young women are related to dowry (Huda 2006), and pressure may continue after marriage, as the groom's family may make ongoing demands of the bride's family (Parveen and Leonhäuser 2004); including in dowry-free love marriages (Naved *et al.* 2001). Payment of dowry may also result in daughters not receiving any inheritance from the family property (Huda 2006). Therefore dowry establishes and consolidates patriarchal norms, compromises women's rights and isolates girls from their family of origin (Chowdhury 2010; Huda 2006; Lindenbaum 1981). As one participant in our research notes:

> Problems are mostly around marriage and dowry. Say if there had been no dowry in the country, we would live more peacefully.
>
> (Older man, Barguna)

Divorce

Divorce is uncommon in Bangladesh, with only 2 per cent of women registered as divorced (NIPORT *et al.* 2005). The low divorce rate may be associated with poor statutory protection of divorced women and the perceived societal shame associated with divorce. Where there is a divorce, the reasons given include the husband's extramarital affairs, problems with dowry and problems between the families of the bride and groom, along with men marrying additional wives to gain more free labour and first wives separating from their husband due to poor economic support (Bhuiya *et al.* 2005; Zaman 1995). Marriage laws give husbands the right to unilateral divorce but they also give women some rights to child custody, financial compensation and maintenance (Schuler *et al.* 2008).

Women are significantly socially and economically disadvantaged when divorced, separated or deserted. Upon divorce, the mother has custody rights over sons until the age of 7 and daughters until menarche, at which time the father can demand that the children be transferred to him. Regardless of who has custody, the father is the legal guardian and must pay maintenance, in addition to dower (the amount specified at the time of marriage to be paid to the women on divorce); however, this payment is not obligatory (see Gierbo and Imam 2006; Huda 2006).

Remarriage causes a woman to lose her rights to dower, child custody and maintenance (Schuler *et al.* 2008). Women are more vulnerable when they relinquish this right, or their husband does not have the finances to fulfil his obligations, and women lack access to legal remedies to address these anomalies (Huda 2006).

Following divorce, if a woman is unable to stay with her husband's family or family of origin, they often migrate to the city for work (Ahmed and Bould 2004). Most women experience financial difficulties in supporting themselves and their families, with minimal prospects for remarriage due to high dowry demand for divorced women (Bhuiya *et al.* 2005). Daughters of divorced mothers have less stable marriages and are more likely to experience divorce or desertion themselves (Amin 1997).

The family household

The traditional Bangladeshi family household structure illustrates the institutional oppression of women. Upon marriage, women relocate to their husbands' village and household, where they are expected to obey their husbands and families-in-law (Blanchett 2001; Zaman 1996; Cain 1978). Culturally, a man is considered superior, and woman are viewed as inferior to their elderly in-laws (Zaman 1996: 65). In these circumstances, without the support of their family or village, young women are particularly vulnerable. Older women are often active participants in maintaining the oppression of their daughters-in-law (Chowdhury 2009: 605; see also Cain *et al.* 1979). Culturally, women are expected to bear children, which gives them bargaining power with their husbands and families-in-law (Chowdhury 2009). There is an overriding cultural preference for sons in Bangladesh (Chowdhury 2009; Jayaraman *et al.* 2008), because 'sons are the key to security in old age' (Ahmed and Bould 2004: 1334).

However, these attitudes are changing, and women working for wages do not believe it is absolutely necessary to have sons (Ahmed and Bould 2004). Traditional gender roles ensure that women are responsible for childcare, including making decisions regarding children's health care and nutrition. This can involve selective deprivation of girls in favour of sons (Hossain *et al.* 2006). Indeed, one in six female infant deaths in India, Pakistan and Bangladesh are due to neglect and nutritional discrimination (Mehta 2004).

Women generally have limited decision-making power, including limited financial control (Chowdhury 2010; Ahmed 2004; Kibria 1995; Kamrunnahar 2007). Because of *purdah*, women also have extremely limited mobility outside the home. A spatial mobility map in Parveen and Leonhäuser's (2004) study showed that women's movement was limited to the natal family home, the local town to buy clothes, and the health centre. Due, in part, to the confines of *purdah*, it is rare for rural women to have personal friends outside the *bari* (family compound) (Pitt *et al.* 2003).

Some women are heads of households – 13 per cent of rural households and 11 per cent of urban households are described as female-headed households (FHH) (NIPORT *et al.* 2009). This occurs through widowhood, divorce or desertion by husbands, long-term migration of male members (often for employment) and men's loss of employment capacity (Mannan 2000; Amin 1997; Alma 1985). This is not necessarily viewed as emancipation for women – rather it is seen as a continued social abandonment of women by husbands, fathers, brothers and uncles who cannot provide for them.

In 2000, nearly half (49.1 per cent) of FHH were in absolute poverty compared with 44.1 per cent of male-headed households (MHH). Additionally 26.4 per cent of FHH live in ultra poverty compared with 19.6 per cent of MHH (BBS 2009). FHH are also discriminated against through prohibition from participating in *shalish* (community council), wedding negotiations or performing marketing activities in public; they have a decreased likelihood of paid work than MHH; less earnings than MHH; and less familial access to health care and education (Mannan 2000).

FHHs have less control over land resources; greater dependency on wage income; higher involuntary unemployment; lower education/literacy level of the head of household; higher age of the head, often greater than working age; and a smaller amount of available household labour (Mannan 2000: 30). They also struggle to maintain the family household and encounter difficulties defending their right to their property.

Women's activism

Despite or because of the subordinate position of Bangladeshi women, activists, professional women and scholars lead a strong women's movement with a long history in Bangladesh. The Constitution provides the reference point for women activists as it encompasses gender equality and provides for affirmative action for women through the provision of the reserved seats for women in parliament.

The women's movement began in the early years of the country when the rehabilitation of the women violated in war was a critical issue. The government established a foundation that supported a number of women's groups. This coincided with the UN International Year of Women and the subsequent Decade for Women. These milestones assisted the development of the women's movement in Bangladesh, and groups set about documenting the status of women. This information was presented to the Mexico International Women's Conference in 1975. Issues that were taken up by the movement included the recognition of women's work, women's legal status, education access and health needs. The women's groups were documenting these issues and were working in concert with government officials who were being exposed to international forums discussing women's rights. As one activist described it, 'their commitment, it was our strength'. In this environment the women's movement continued to expand, starting with urban areas and moving out to the villages along with NGOs working with microcredit and other women's empowerment programmes.

The assassination of Prime Minister Mujibur Rahman in 1975, and the instigation of military rule, changed the relationship between government and the women's movement. The story of the women's movement is related here by one prominent activist who participated in this research:

> At that time, the military rule wanted to legalise their existence, so they wanted support from us. It was the late seventies … we were under pressure from the then government to give a statement in favour of them, a political statement. At that time I was the president of the Organisation for Women. I was very strong and said, 'No, we can't change that statement. We are not political.' Then she, the Minister for Women's Affairs, said 'No, this is not a political statement, don't you want to go for progress of the country, peace of the country?' I replied, 'Yes, we want that, but our way of thinking and your way of thinking is different, so please don't involve us in this thing.'

The slow incorporation of religion into the affairs of the state by the military rulers and the adoption of Islam as the official state religion changed the national commitment to gender equality:

> And then we protested that, the women's movement protested it very much, even with our voices stronger than theirs.

The women's movement changed tactics, introducing lobbying and dialogue with government and successfully requesting that government ensure gender training for bureaucrats. This training was developed and delivered by women's movement activists and focused on gender awareness and women's rights. Initially there was scepticism and outright hostility, but they persevered. The movement also lobbied Dhaka University to establish a Women's Studies programme, a project that took 14 years but culminated in the establishment of the Women's Studies department in 2000.

The women's movement worked closely with the more progressive government elected in 1996:

> We interacted with them very closely and they adopted a National Policy for Women. It was very progressive, and it was the result of the women's movement. And by that time the [Beijing conference] had taken place and immediately after [Beijing] the government formed a core group to work on implementing [Beijing]. I was one of them, they invited six of us – two from our women's group and four from other organisations. So, we helped the government for two years to develop a National Action Plan.

The women's movement was very active in lobbying government to introduce gender mainstreaming in the post-Beijing environment.

> We started lobbying and we developed the concept of mainstreaming gender. So, we first educated ourselves about the planning process, about the country's five-year plan. Then we formulated a working paper on mainstreaming women with all the strong justification about inclusion in the fourth five-year plan. It was in the year 1990 – we went to the Planning Commission, to the Planning Minister, and we handed it over to him. And at the same time, we held a press conference and interacted with other activist groups … we succeeded in including a gender-mainstreaming approach.

The movement for women's rights was under significant pressure during this period. Our informant describes it as follows:

> Somewhere around 2003 or 2004, the government in power changed [the women's policy], without any consultation, whereas the '97 government had huge consultations with women's groups. So, then again, we came out onto the streets protesting, and nothing happened and then we had the caretaker government with the army. They said, 'Okay, we'll declare this policy again as adopted.' So in 2008, there was a huge backlash from the Mullahs and Imams, because they were saying that the policy is contrary to Islamic laws.

The contest over women's rights continues. As the passing of the 2011 amendments to the women's policy demonstrates, fundamentalist Islamic leaders have led violent demonstrations and strikes against gender equality measures. The area of women's advancement is a volatile one in Bangladesh.

The Bangladesh women's movement continues its active campaigning and has taken its fight to the international arena. It is part of South Asia Women's Watch and of the Asia-Pacific Women's Watch. Through these bodies they work to progress their demands through the Commission for the Status of Women (CSW):

> We have a forum. It consists of more than 600 organisations, national and local levels. And every year before the CSW, we hold a national consultation. First we hold several small-scale consultations, then a big one at national level. We also prepare a report for CSW. We involve as many organisations as possible, we identify the team, and we allocate the team to different organisations, between groups, according to the relevance.
>
> Then in South Asia there are five sub regions, East Asia, Central Asia, South East Asia and Pacific and these are all involved in the Asia-Pacific Women's Watch. From the Asia-Pacific we take the report to the CSW. This way we take our voice, local to global.

Despite the active engagement of the women's movement and its dedicated leaders, there is a continued threat to women's rights from religious fundamentalism. There is some danger in speaking out as social commentator and activist, Taslima Nasrin has discovered. She has been forced into exile for her forthright views on fundamentalism and the rights of women.

The conflicting principles between a state-sanctioned Islamic religion, constitutional rights, a market economy and an international women's rights movement has led progressives to oppose religious fundamentalism and to demand secular democracy (Hossain 2007). Nonetheless, Khushi Kabir, former head of ADAB, reminds Western scholars and commentators that fundamentalism is not the province of Islamic states – religion has been used across the world and across time and space to suppress women (*Green Left Weekly* 2006). In discussing Taslima, Kabir notes:

> The Western media portrayed Taslima as the victim of an oppressive society where women are cloaked in silence. They gave the impression that as soon as anyone resists fundamentalists, they are crushed. This denies the reality where hundreds of women, every day in their ordinary lives, are confronting fundamentalism and resisting it. When you are an individual like Taslima fighting the fundamentalists, it is a lost cause. The only effective way to resist and survive in a country like Bangladesh is to be part of a movement. The women in the villages can fight the mullahs and landlords because they are united and have the numbers.
>
> (*Green Left Weekly* 2006)

This unity of women and the women's movement at national and local levels has resulted in constant activism for women's rights since the Liberation War and the establishment of the Bangladeshi nation. In the early years activists such as Nilima Ibrahim, Bashanti Guhathakurta, Naushaba Sharafi and many others gave support to the birangonas, demanding the recognition of the rights of women and the establishment of the War Crimes Tribunal for collaborators.

Yet increasing religious fundamentalism, the rise of Madrassah schools and the conservative backlash in politics have led to an increasing encroachment on women's rights since the 1980s. Nari Pokkhyo, a small women's group, unsuccessfully challenged the attack on women's rights in the High Court during this period. Ironically, fundamentalist opposition to gender equality measures occurred at the same time as an emerging market economy was increasingly demanding women's cheap labour in garment factories, shrimp-farming and other employment; and the introduction of family planning measures aided women's entry to the workforce.

Violence against women became politicized in the 1980s and activists, aided by international NGOs, advanced a human rights perspective. The women's organization, Mahila Parishad, demanded action on Islamic personal laws and greater political representation of women. A Charter of Women's Rights was developed in 1990 by activists and led to the establishment of Oikkyo Boddho Nari Samaj, a group championing women's rights.

The Bangladesh Nari Mukti Sangsad (Bangladesh Women's Liberation Council) was established in 1980 as part of the Workers' Party Bangladesh and has continually advocated women's rights and the implementation of women's policy. In 2008, when the women's policy was being reviewed and a committee of Islamic scholars led a movement against women's rights, the convenor of the Women's Liberation Council, Hazera Sultana, stated:

> The fundamentalist and communal fanatics in the country are whipping up tension and agitation … we watched with abhorrence as the government yielded to the fanatics' hollow intimidations … the policy professed to remove all kinds of disparities affecting women's participation … who else but the hypocritical religious fanatics upholding outdated medieval dogma should oppose this?
>
> (*Bangladesh News* 2008)

Protests by women activists continue. In 2013 for instance, Nari Mukti Sangsad staged a protest against a fundamentalist mufti who they said had been continuously condemning women in his sermons. He has called for women's access to education and employment to be restricted, for women to adopt strict *purdah* and for them to be denied access to family planning. Several leading activists and academics signed a communiqué condemning his 'indecent' and 'perverse mentality' (Mustafa 2013).

A Social Resistance Committee (SAC) was also formed in 2013 by 68 NGOs, calling on the national government to resist fundamentalists and to actively advance the women's policy. Sufia Akhtar, a spokesperson for both Mahila Parishad and SAC, demanded the government resist the fanatical attempt to turn secular Bangladesh into a 'Taliban-style pariah state with the policy of subjugating women' (Chowdury 2013). The threat of Islamic fundamentalism to women's rights is constant, and the national women's movement is strong and responsive, although under-resourced, disparate and constantly under threat.

At local levels, women's movements are also common across Bangladesh. Traditionally women have been critical to agricultural production, selecting and storing seeds, keeping home gardens, working in the fields and processing crops. New measures introduced to develop commercial agricultural production over the last two decades have eroded women's agricultural centrality. These measures include the building of embankments, the harvesting of forests and the establishment of ponds for shrimp-farming. These changes have often led to localized resistance movements by women. For example, in 1989, a woman farmer, Korunomoyee, who was leading a protest action against the co-option of land for shrimp-farming was murdered by an armed gang. Each year on the anniversary of her death, local village women commemorate her memory as a symbol of resistance to the encroachment of commercial agriculture (Hossain 2006).

Women are very conscious that the development of export-based income reduces their capacity to grow food for their families and reduces their input to that of waged labourers and workers on their own lands, often at the mercy of overseers and gangs, and at increased threat of violence (*Green Left Weekly* 2006).

The oppression of Bangladeshi women is evident from national to local levels. New threats emerge and are challenged. Strong and courageous women lead movements all over the country and at international forums, seeking rights and freedoms for Bangladeshi women.

Summary

Bangladesh officially recognizes gender equality through its participation in international forums and charters, and through the Constitution and national Women's Policy. Additional forces that are aiding women's empowerment include their increasing access to work and education, globalization, rural-to-urban migration and greater acceptance of women's economic contributions. However, the acceptance of Islam as the state religion provides significant tensions between equality and traditional customs that disadvantage women. In this chapter we have indicated that customs surrounding freedom of movement, marriage, dowry, divorce and family structure restrict women's freedom and lead to significant disadvantage.

Fundamental to the advancement of women in Bangladesh is a strong and active women's movement and the civil society organizations (both INGOs and NGOs) that foster and develop programmes that advance gender equality, and support and acknowledge the human rights of women. The need for hyper-vigilance is evident in the violence and protests by religious fundamentalists that follow amendments to women's policy and advances to women's rights. Yet the women's movement is active. Bangladeshi women's leaders face constant challenges and new erosions of their rights. Does the emerging threat imposed by climate change signal a further oppression or a new space for women's empowerment?

Ferree (2012: 4) introduces the notion of relational radicalism to suggest that ideas that may appear radical in a particular society at a particular time and place can in other circumstances seem quite tame. Thus current circumstances in countries such as Bangladesh are not fixed or unchangeable and may be open to challenge in the context of climate change – *abnormal times* when the taken-for-granted is exposed. What once appeared radical and unattainable becomes possible. What climate disasters may expose are the taken-for-granted injustices built into the system of governance in Bangladesh.

4 Existing circumstances

Shaping a life in rural Bangladesh

This chapter focuses on the lives and livelihoods, customs and practices of the people in the Bangladeshi villages who took part in this research. In describing the demographic characteristics and daily practices of the participants, it brings to life the everyday/every night worlds of women and men living in rural areas of this rapidly developing country. The descriptions, quotes and data reproduced in this chapter give an indication of gender divisions, workloads and traditions, and provide background to the material in the following chapters that describe the impacts of climate events on already changing, but highly gendered ways of life.

The data reveals the almost universal acceptance of gender roles in the expectations, actions and ongoing activities of the people who contributed to this research. Gendered patterns are evident in the division of household labour, ownership of land and resources, and agricultural and paid work. A dominant masculine hegemony results in men owning the majority of assets and dominating decision-making, and women undertaking work in the fields and moving into paid labour.

Quite apart from the impact of climate change and catastrophic events, there have been quite dramatic processes of restructuring in these rural areas underway in Bangladesh as there is in other parts of the world. The exposure of rural villagers to a market-based economy has led to a steady erosion of well-defined and separate gender roles, and the lines between what has been traditionally defined as women's and men's work are blurring. Women are taking greater responsibility for agricultural labour and food production for the market and for home consumption.

The need for income generation affects all members of the family and is resulting in significant changes in labour patterns – the greater out-migration of men, the movement of many women into the workforce and the need for young people to work from a young age. What has not changed is the responsibility for household labour – this remains women's responsibility.

Assets

Ownership of assets reveals the gender structures (see Table 4.1). Nearly half of the 610 quantitative survey participants own livestock, predominantly

Table 4.1 Ownership of assets

Asset	*Women*	*Men*	*Total*
Don't own any	39%	29%	34%
Livestock	57%	34%	45%
House	2%	11%	11%
Homestead land	9%	44%	27%
Pond	3%	20%	12%
Farming land	1%	35%	18.5%
Vehicle	0.5%	20%	10%

hens, cows, ducks and goats – very few owning more than a handful. One-sixth of the participants own some land, but approximately three quarters do not own the land on which their house stands, although this disguises the fact that nearly half the men own their homestead land as opposed to very few women who traditionally will move to their husband's house on marriage. Only 1 percent of the women who took part in this research own any land at all. One-tenth of respondents have some ownership of a pond for household water, cooking and bathing, and one tenth own a vehicle – most often a bicycle, rickshaw or motorbike. One-third do not own any asset (this is higher for women). Where there are assets, men are far more likely to own these – although women are more likely to own and rear the usually small number of livestock including hens, ducks and goats. Men own the majority of assets apart from livestock, ensuring that women are highly dependent on their husbands and family for their livelihoods and security.

Decision-making

There are also very significant gender differences in decision-making activities. Participants were asked about various decisions taken within the family – if decisions are made in certain areas, who is ultimately responsible for these? Almost all reported that men make most decisions. Over 70 percent of women report that they make no decisions at all and this includes on issues relating to livelihood strategies, health, children's education, disaster-planning, post-disaster strategies and, surprisingly, family-planning. Table 4.2 outlines the responses for a variety of categories. While not all participants responded to every category, the table demonstrates the gendered nature of decision-making within households.

Nonetheless, in focus groups, women report that this is changing as work roles evolve and as women are earning their own income. This has given women more power and control and has led to some women now going to the market themselves, and unescorted. Women in a focus group in Barguna made the point that the change is not due to women 'avoiding the veil',

Table 4.2 Areas where respondents felt they could make autonomous decisions

Decision-making area	*Women*	*Men*	*Total*	*Comment*
No decisions	71%	0%	34%	
Livelihood strategies	5%	100%	54%	
Own health	29%	29%	29%	
Division of labour	6%	43%	25%	
Own paid work	6%	30%	18%	
Family's health	5%	24%	18%	
Children's education	7%	22.5%	15%	
What to do in a disaster	4%	17%	10.5%	
Family planning	4%	16%	10%	

meaning they are still careful, although the practice of *purdah* is dying out in these regions. Rather it is due to the new economic contributions that women are making to the household, a contribution that allows the family to have more nutritious food to eat. This ability to provide food has given women significant new power within their relationships:

> [Where a woman earns more than a man it creates] a crisis. Women are rejecting their husbands' [traditional contribution to] daily activities such as buying things from local market .... Women are coming to local markets more.
>
> (Focus Group, women Barguna)

> Now the ball is in the court of the women such as the wife or the daughter-in-law. The father has become old and his pocket has been empty ... (sound of laughter). In older days, the income was less. This is important – rather than only blaming the women for avoiding the veil. ... But [previously] the father was the only bread winner. Say every day he could earn 2 kilos of rice. It barely met the family's need. They had to eat water with a very little amount of rice for everyone and they have to eat it with green chilli. Now they do not need to 'eat' water. That the veil is avoided is true, but it works better. Now that the male adults and the wife work, they can even afford 1 kilo of milk.
>
> (Focus Group, women Barguna)

Household work

Some traditions are more intractable than others. Women and men still view household work as women's responsibility and men do little or no tasks within the household compound. Housework in Bangladesh includes gathering fuel for cooking and water for cooking and bathing, food preparation and much more. In many isolated areas women also do home schooling with their children, keep a home garden, ensure the maintenance of their homes, livestock pens and stoves, and lay down feed for the rainy season. As one older woman from a Char area of Gaibanda explained:

> This is Char area … The women have to work a lot at home. They have to water the plants. They need to give food and water to the cow and goats. They have to repair the cottage. There is a lot of work. Men think that the women do not have a [heavy] workload. Coming from the field they shout to their wives about the pending household work. They do not understand.
>
> I wake up at early morning. Sweep the house and yard. I have to feed the cattle then. I need to cook for the workers who are hired. I need to send the food to the field. In between that I sit with my children and teach them. That is not so effective; I need to get them a tutor! I do all this work all day long. At noon I have to cook lunch. I take a bath along with the children. Again in the afternoon I sit with them. After Zohr prayer I need to cook again for dinner.
>
> Suppose the rainy season is going to start. I should look at the roof of my cottage. Maybe I need to elevate my house … make another stove. The crops may be damaged and I need to keep those safe. I make the land higher and sow the vegetables. …During rain snakes and insects may bite my children. My children are little and I have to keep my eye on them, so that they cannot go to the river. They may fall in the river.

Across the world women complain about the invisibility of their work and the lack of acknowledgement of their efforts – the women of Bangladesh are no different, albeit their household tasks are more arduous and complex. Women in a focus group in Barguna complained about this invisibility but also about the implications of marriage for women's time and labour. They also gave some insight into their personal relations with their husbands.

> But our [labours] are unseen. According to them [the men] they work and earn and we do not, just cook and eat sitting idly inside the house. But in reality women start working early in the morning after waking [and work] until sleeping. No one counts our work. Men believe that women's only duty is to do household work. There is only one thing we have to do more than them – that is to mandatorily sleep with them. (Laughter) …. But this is extra work for women. In this case only the husband has the right to be happy.

The significance of household work, family and customary practices, as understood by men, was also explained to us in a male focus group in Barguna. Men note that it is they who can go to the markets, not their wives, but it is the women who must cook. A good wife is viewed as one who looks after the husband's family well and preserves the honour of the family:

> If my wife wishes to take care of my relatives, she can but I cannot. I can go to the market and get the ingredients … but cooking depends on the wife … If the wife is good [my relatives] get good treatment.

Despite any additional work undertaken by women in the fields or to secure income, their responsibility for household work remains unchanged. Some women suggested that this division is innate and that men cannot do these tasks for physical rather than cultural reasons:

> No, [men do not cook] they do not do that, because it is our mental set up. This is suitable for only women, isn't it?
>
> (Interview, female, Barguna)

Men and work

Men work in the fields, as day labourers, fishermen, shopkeepers, and transport workers. The precarious nature of agricultural production, particularly as climate variability reduces yields and returns, forces many men to take alternative income-generating activity. For many, this necessitates migrating away to work in the cities as construction workers or day labourers. This out-migration creates significant tensions as men are very aware that this may leave families vulnerable and their wives without a male escort. Further, many men describe the work they do far from home as hard, lonely and isolating. Yet increasingly families are reliant on this remittance income generated elsewhere. However, the cost to all members of the family is very high:

> Sometimes we [migrate for work], sometimes we do not. There are so many things to worry about. If we leave then our children and wives stay home on their own.
>
> (Older man, Gaibandha)

Women also worry about their sons who leave home for work, often far from home in cities with little family or other support. These young men and their mothers know there is no alternative if they are to support their families – but that does not reduce the hardship and isolation:

> Here there is no work. I have two sons. This year one went to work as a day labourer. Another son has been doing this for last 3–4 years. With this my family runs in a poor condition … The younger one worked at a stall. There he got only food not money. He started to work as a labourer for last 25 days. He digs soil. But cannot send any money till now. I

> called him yesterday. He cried out and said, 'This job is too hard'. He cannot stand it.
>
> (Older woman, Gaibandha)

Men whose agricultural and fishing work is seasonal may migrate for parts of the year only. In order to ensure that the family has adequate food, it is necessary to source additional income during these periods of low activity. Fishermen, for example, are amongst those who must find alternative sources of income for parts of the year. Most will migrate to the cities where they are employed in construction work in the rapidly developing urban areas, or work as rickshaw drivers, stallholders or day labourers. Poverty and financial hardship force others take on more dangerous occupations in order to stay closer to home. In Satkhira, for example, some men brave working in the jungle catching crabs and other sources of food. This is extremely dangerous as they are exposed to attack by tigers, crocodiles or snakes, and being kidnapped by pirates:

> I go fishing – but not throughout the year as there is a certain fishing season. [I catch] little tiger prawns and mixed fish. We catch them six months of the year. [For the rest of the year] some people go to the jungle for crab hunting. Some work as day labourers. But it is difficult to go to the jungle for fear of the pirates. The amount of ransom is too high.
>
> (Older man, Satkhira)

> [My husband] can work at the shrimp farms as a labourer – otherwise we will have to go to the jungle for work. There are [great] risks from tigers and snakes. Even a few days back a 22-year-old boy died. [Another woman's] husband was also killed by a tiger. He went to the jungle in search of leaves with his son when the tiger attacked and killed him. In another incident, two sons and their father were all killed by a tiger. So it is safer to be at home eating less than to go to the jungle. People earn up to 10,000tk in a week from the jungle, but we discourage it. Previously people used to work seasonally in the jungle for about 3 months, but now they work there round the year because there is no work here in the village, no paddy in the fields.
>
> (Older woman Satkhira)

Women and paid work

Younger women now have a way to avoid this subordination. The opening of the garment factories in Dhaka has given them an acceptable alternative – one that provides remittance income for their families and allows them to escape the threat of early marriage and a lifetime of servitude. Many younger women, and some young men, migrate away for work, either in garment factories or as domestic workers. Nonetheless the issue of their safety and hence the 'honour' of the family that is bound up in women's behaviour and appearance does not escape them. For the girls themselves the threat of violence and poor work conditions is very real:

> Let me tell you about [my daughter]. Her father works as a labourer in shrimp farms. She used to work as a housemaid in households at Dhaka. She would get beaten at times. I taught her tailoring so that she can work here once she grows up. [When I was a housemaid] they used to beat me up too. At times they would not let me eat.
>
> (Older woman, Satkhira)

Poverty and straitened financial circumstances are causing a revolution in women's working lives. While marriage defines these roles for women and places them in a subordinate position to their husbands and their husbands' families, their daily lives have become extremely busy with the addition of paid work, livestock-rearing, work in the fields and household work. Many now work outside the home, breaking down old customs and practices that confined women to household compounds. For a variety of reasons including economic restructuring, economic development, financial crisis and the growth of export industries, women are now more likely than any previous generations to work outside the home, and this both challenges conventional customs and raises questions about 'honour' and male control. There is a significant change occurring in women's livelihood strategies, a change that puzzles and upsets many of the older men. Women view the need to work for income as a necessity to support their family, a necessity that nonetheless is challenging customs that formerly constrained their lives. Yet the low pay received and the harsh conditions under which women work suggests that many women may be replacing one oppressive environment for another:

> Five years ago women would not come out of their houses. But now they are forced to come out.
>
> (Older man, Satkhira)

> We are poor so we have to sell our labour. We all are forced to work in the fields due to financial crisis .... [We earn] 50tk daily [this is approximately 70 cents in Australian currency].
>
> (Women's focus discussion group, Satkhira)

Like their counterparts in developed countries, families in Bangladesh now require more than one income to survive. Because women's wages are so low they are more likely to be employed locally as day labourers as they are 'cheaper' than men. There is also some suggestion that women are more likely to obtain local employment after a disaster in reconstruction projects because they are paid such low wages and are seen as more employable. This has several implications including higher employment opportunities for women, albeit under oppressive conditions, and the greater impetus for men to leave to secure income elsewhere:

> It is true that it has become tough to maintain a family with the income of just one member, on the other hand, it is also true that those who hire labourers, they are required to pay higher wages for male labourers. That is why people now prefer to hire women workers.
>
> (Focus group, teachers, Barguna)

> [After the cyclone] women worked more than the men … we cut mud and dug ponds. We did those things in return for money. We did everything … look if we want to eat we must work.
>
> (Female focus group, Barguna)

The exploitation of women in construction work, through lower levels of pay, also involves them working in unsafe conditions. Women consider that employers give a very low priority to their well-being and safety. While women working as day labourers express concern about these conditions, they also note the breaking of customs and the conflict between the need to work and to maintain their 'honour'. Most women resolve this through modest dress:

> I am working on road construction. I cover myself properly. So, I don't face any problem.
>
> (Older woman Gaibanda)

Women whose husbands are farmers also report that they are now more likely to be working in the fields with their husbands as their families are less able to afford paid labour. When women work in the fields there is a division of labour with women doing the weeding and planting and fetching:

> I weed the crop field. I help my husband by taking the ploughing materials. During the time of corn I sow the seed in the ground by hand.
>
> (Young woman, Gaibanda)

> Yes, we've got to go to the field following our husbands. If the family is in crisis, my husband as well as me has got to face that situation. If we both work together, only then we can raise our children properly. If one is working in a family and the rest are not helping, that won't work out. Then rearing the cattle can help. If my husband is going out and bringing money, I got to help him.
>
> (Young woman, Gaibandha)

Women who are widowed have no option but to work and will do whatever is available. Their lack of a protector is keenly felt and they feel the need to take measures to preserve their honour. As one woman from Gaibanda explained:

> After my husband's death I had to suffer a lot. I started to work at the other homes. I pulled the nets in the river. I gathered pulse and chillies from the fields. My eldest son worked for 20tk. My mother helped me many times. In fact I am leading my life under the shadow of her. I have not lost my respect and I worked only among these neighbouring homes.

The movement of women into the paid workforce matches a worldwide trend. For the women of Bangladesh this trend is marred by unsafe conditions including the threat of violence. Their paid work is further constrained by low wages, poor access to work, customary expectations about household work and maintaining their reputation or 'honour'. Critical to their working conditions is their level of education, as, the higher their achievement, the more likely they will be employed in better-paid and safer jobs.

Education

More than a third of the 610 quantitative survey respondents has no formal education (131 women and 109 men); 57 (9 per cent) had not completed primary education (22 women and 35 men); 66 (11 per cent) had completed primary (30 women and 36 men); 108 (18 per cent) had some secondary (57 women and 51 men); 47 (8 per cent) had completed secondary (18 women and 29 men) and 52 (8.5 per cent) had a higher degree (24 women and 28 men). (See Figure 4.1.)

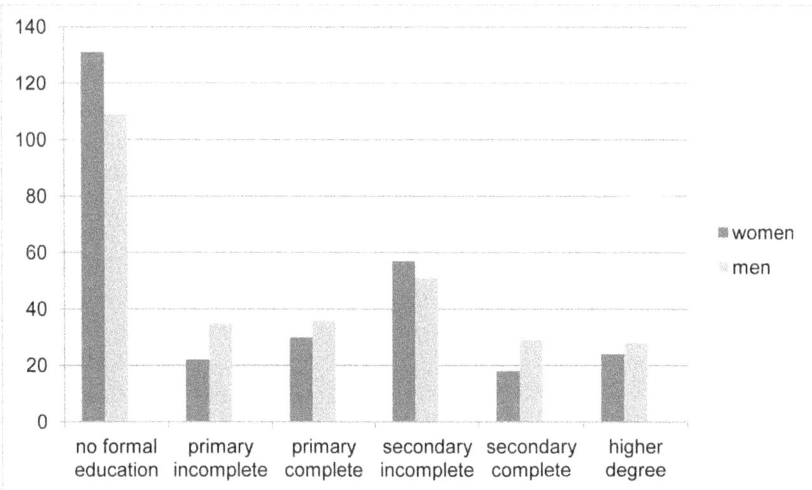

Figure 4.1 Educational attainment

Despite education providing increased work opportunities, male participants in focus groups express some ambivalence about education for women, although women have no such reticence. Men also view education as a problem because educated women may not wish to do housework, and men perceive this as a potential source of family tension. By contrast they also note that education allows women to work in higher paid jobs and therefore protects them from early marriage and relieves the family of the need to provide dowry payments:

> When it comes to cooking, [the educated woman] disavows and asks her husband to appoint someone to cook since she gets tired due to her job. These families are full of problems. All the educated families are packed with problems. Outside is full of knowledge but inside is totally empty … But education helps to avoid dowry.
>
> (Male focus group, Barguna)

Nonetheless most participants ultimately note the importance of education. Women are especially motivated to give their daughters as high an education as is possible. For many women this means they go without essentials to ensure that their daughters remain in school. One woman notes that this includes her going hungry in order to afford her transport to school:

> My husband is a day labourer. I'm educating my children. One daughter is studying in [high school]. We remain hungry but I had to give money for the fare for [my daughter to catch the school bus] to Barguna. Every day she needed 40–50 taka but my husband couldn't earn even 100 taka. Then how can I eat? I eat rice with chilli paste but give my daughter the bus fare. If I buy fish with 40 taka then I wouldn't be able to pay my daughter's education.
>
> (Older woman in Barguna)

The dominant view of education amongst younger women is that it is prized. Those who have the opportunity to remain in school note its value, as the benefits for their future lives are enormous. Those who had been subject to early marriage mourned the loss of their education, with some claiming this loss of educational opportunity meant their lives were effectively over.

The aspirations concerning education have changed markedly in a generation, due in part to government policies including the provision of a free lunch and biscuit and an NGO stipend for girls to attend school. These policies have been enormously effective in allowing girls to remain in school as they relieve the family of the burden of providing food and give them a small additional source of income:

> In the past people were ignorant, there was no NGO to motivate them. But now, everybody motivates us. People are educated now. Children of the ignorant people are also going to school now.

> You have to study. My father was the local chairman. He could sign only. My mother was ignorant. But they educated all of us. We are 8 brothers and 8 sisters. All of my brothers passed IA or BA. We the daughters can read the Holy Quran in Arabic and Bengali. All are educated in my family.
>
> (Woman in her thirties, Gaibandha)

Nonetheless because of the geography of Bangladesh, access to schools remains a problem in many areas, made worse when climate events erode infrastructure, destroy roads and make rivers impassable. There is also growing resistance to the education of girls, particularly in areas where religious fundamentalism has become dominant. We were told in Satkhira that there had been a fatwa placed by local religious fundamentalists on girls who were attempting to go to high school, severely restricting their access. This fatwa meant that girls were subject to violence should they attempt to go to school. Such evidence of growing fundamentalist influence is deeply disturbing for young women seeking to pursue their education.

Access is also restricted where there are no high schools, such as on the Char islands and where transport to the nearest high school is problematic. Many of the Char areas have no high school and Char Islanders have to cross the river by boat to reach schools. This often leads to families withdrawing their sons and daughters as transport is expensive and they feel their daughters' honour may be compromised. Where there is no high school, young girls are more likely to marry at a very young age:

> We have no high school. Poor people cannot imagine that after primary completion they will send their children to Gaibandha. Some people don't want to continue their study by crossing the river daily.
>
> (Young male focus group, Gaibandha)

> School and college is a big problem here. There is no high school, understood? (My daughter) goes to high school which is at Kundepara. She does not get the boat during flood. Transport is a big problem. I have to pay 150 to 200 taka as boat rent every month.
>
> (Older woman, Gaibandha)

Poverty

Notwithstanding the addition of paid work by women, poverty is still widespread in the villages. Key informants from NGO organizations note that there are layers of poverty. The category of 'the poor' is so large in Bangladesh that government and NGOs differentiate the 'poor' into different segments, or classes, including the ultra-poor, the poor and the slightly better off:

> We'll take a step back. We work with multi tiers of poverty. Our job is basically poverty alleviation. So, we use multi tiers of poverty; that means

> the ultra-poor, and the poor; [so we are] able to clarify exactly who they are – then the poor, then the slightly better off amongst the poor. Then for certain programmes, we take the entire community. Say if you are talking about public health, you can't isolate just the ultra-poor or the poor. So, access to financial services is basically for the ultra-poor and the poor.
>
> We have a specific programme, it's called Especially Targeting the Ultra-Poor. And in this Ultra-Poor Programme, we have a definition and this definition was developed through research and experimentation. We use at least four indicators to define ultra-poor and ultra-poor houses, for example, cultivatable land of less than 10 decimals per household. If the household head is a female and the household has at least one member. Whether there is an able member to carry out income generating activities.
>
> (NGO Key Informant)

These distinctions are useful, as it makes clear that many female-headed households are amongst the ultra-poor. This can result from out-migration of males, the death of husbands, low levels of education of women and/or lack of access to employment.

Loans

One dominant strategy used to address household poverty is to take loans – a practice that is increasing amongst the families who took part in this research. A majority of families borrow money as a direct result of their changed circumstances, including economic hardship and climate uncertainty. Disturbingly, half of these families require the loans to buy food (Nasreen 2008; del Ninno and Borsch 2003). A growing trend amongst Bangladeshi families is the use of microcredit, usually given to women for small business development. Because of high levels of poverty and loan debt repayments, NGOs who provide microcredit have become de-facto relief agencies and are providing valued rehabilitation assistance following disasters. When NGOs reschedule micro-finance repayments following a natural disaster, there is a decreased probability that women will skip meals (Shoji 2010).

Many Bangladeshis are now reliant on these loans to manage their livelihoods – more than half of the participants of our survey had a loan at the time of the research. Commercial loans are slightly more likely to be taken out by men and are borrowed from an NGO or bank. Women are more likely to borrow from family, neighbours or women's loans/microcredit schemes. Loans are taken most often for cropping inputs, but also for businesses, house repairs, health treatments, children's education, to buy food and other family expenses. Following the cyclones in the south, women were also more likely to be given access to microcredit and to be employed in reconstruction:

> Women got more help [after the cyclones]. Yes, yes, women worked more than the men. We have done varieties of work. Such as – we worked on

> this road, we cut the mud and dug ponds again. We did those things in return for money. We did everything properly. Look if we want to eat we must work. Our women union member was always with us so that people did not do any harm or teased or misconducted towards us. Our salary has been paid after the end of the week on the basis of per day wages taka 100 or taka 120 (slightly more than 1 USD per day).
>
> (Focus Group, female, Barguna)

Nonetheless there are instances of gender bias in loans and grant schemes, particularly those that are given only to landholders – as these are directed almost exclusively to men who are the nominal owners. Yet, as some NGO representatives note, the out-migration of men results in more women undertaking agricultural work. Nonetheless they are not given access to these generous agricultural schemes because they are not the identified landowner. While women's NGO representatives are challenging these schemes, it is the link between ownership of land and grants that is problematic, as women are not being recognized for their agricultural production:

> Three weeks ago we challenged a very, very well-known national programme, where every farmer is given a card … with that card they can ask for national incentives, subsidy, even monetary incentives, to purchase fertiliser, fuel and all that. They can have a bank loan, credit and everything. [But it is] only for the male farmers. And then, we found out in the last decade [that] Bangladesh was growing at a rate of 5.8 per cent, an increasing number of males are going away from agriculture and an increasing number of females – in five years, four million of them, entered agriculture. They're replacing the male and still, they're not recognized. We talked to the Extension Service, who identifies farmers. [They implied] farmer means male because, there is a lack of ownership of land [by women]. The other thing is just mental I think, partly mindset. It's a mindset.
>
> (Focus group, Key informants, NGO)

Food security

A very significant percentage of our participants in the survey (67 percent including 211 women and 198 men) reported they had not had enough to eat often or sometimes during 2012. A further 99 (16 percent) stated that they had enough but not the right kind of food. Only 107 (17.5 percent) stated they had had sufficient food. (See Figure 4.2.)

When asked who eats less when food is limited, both husbands and wives report that women are eating less, although women report this more often than men. Table 4.3 outlines the responses of women and men to this question. There is no doubt that the expectation from both women and men is that, when food is short, women will eat less.

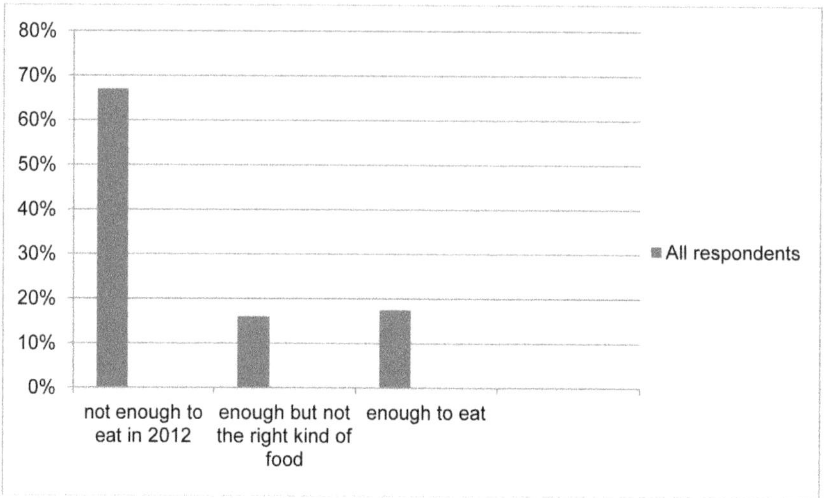

Figure 4.2 Levels of food security in 2012

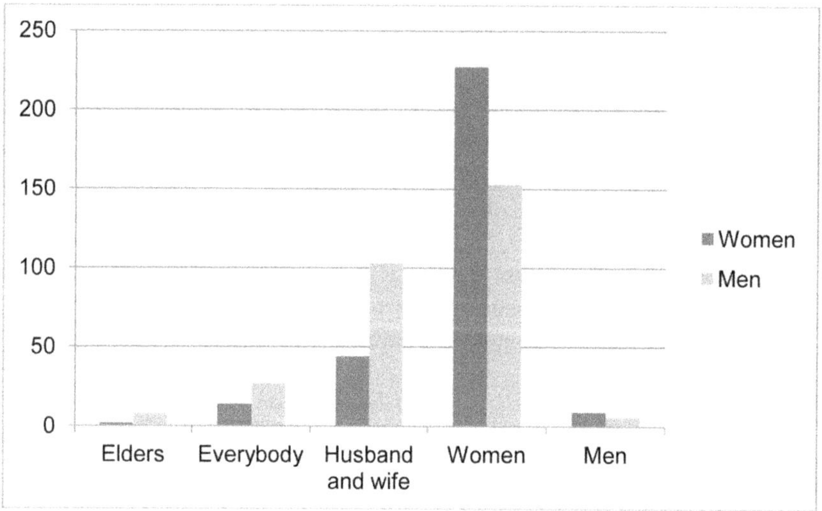

Figure 4.3 Who eats less?

Respondents were asked what were the critical factors that related to their food insecurity and that resulted in many going hungry and undernourished. The dominant issues noted include weather factors, less access to paid work and declining income (see Table 4.4). This information provides insight into the interaction between disasters and other issues of complex restructuring in the rural areas affected. It is the interplay of a number of factors that leads to food insecurity.

Table 4.3 Who eats less?

Who eats less	*Women respond*	*Men respond*
Elders	0.5%	2.5%
Everybody	5%	8.5%
Husband and wife	15%	32%
Women (wife, mother-in-law, daughter)	76%	48%
Men (husband)	3%	16%

Table 4.4 Reasons for food insecurity

Reason for Food Insecurity	*Women*	*Men*	*Total*
Scarcity of work/ unemployment	30.5%	28%	29%
Lower income/less profit	30.5%	23%	27%
Poor crop production/damage	6%	9%	8%
Natural disaster	5%	9%	7%
Increasing cost of food	4%	5%	4.5%
Scarcity of fish	6%	5%	4%
Bad weather	4%	4%	4%
Communication problems during rainy season	2%	2%	2%
Scarcity of food/ lack of stored food	0.5%	2%	1%

When queried on the worst times for food security, respondents indicate seasonal variations in food stores confirming the phenomenon of seasonal hunger. There are periods of the year when people are critically food insecure – these relate to the seasons and production cycles. Over one-third noted March/April to be the worst time, more than half June/July and nearly two-thirds August to November. For up to six months of the year, many people in the rural areas of Bangladesh suffer seasonal hunger and malnourishment and it is during these times that women eat less in order to ensure there is more food for the rest of the family.

One of the outcomes of seasonal food insecurity is that there is now a greater reliance on the market place to purchase food, and many secure loans for this purpose. Almost all respondents report purchasing at least some rice on the open market, and only one-third growing rice for their own needs. Most purchase at least a portion of their fruit and vegetables in the market and approximately one-third are able to provide some fruit and vegetables from their own farm or home garden. Almost all purchase their fish and other meats from the market. (See Table 4.5.)

These figures suggest villagers are less able to grow food for their own needs and are becoming much more reliant on the market. This necessitates income

Table 4.5 Food purchased from the market or grown

Food item	*Purchase*	*Grow some for own use*
Rice	90%	33%
Beans / legumes	90%	13%
Fruit / vegetables	86%	29%
Fish / meat	92%	20% (livestock) +19% (fishing)

generation and/or loans to purchase food, thus facilitating the growing trend to paid work and out-migration.

One NGO informant noted gender differences in the way women and men understand food security:

> When we asked the men, 'What should be in your meal? What is food security? What is the preference?' Males always say, 'For me and family, rice and meat'. Red meat, basically the priciest of all. The women say, 'Well, a mix of dhal, that means lentil soup, and maybe a vegetable could be a good meal for my family.' She's much more practical. Wise and practical. She perceives the world in a practical way. Really, males and females have different perceptions regarding food security, house and food. And the entire concept of well-being is completely different.
>
> (Key informant, NGO)

Gender equality

Our research produces very critical insights into existing and ongoing gender inequalities and the factors that are enhancing or restricting equality. Access to work, the importance of alternative income and the need to adjust to changing economic realities are providing a strong impetus to women's empowerment. One older woman notes the changes that have occurred in her lifetime. When first married, her movement beyond the home was severely restricted; she practiced *purdah* and was beholden to all members of her husband's family:

> If people like you came here in the past we would cover our faces with veils and we would look to the side. Yes, the times have changed. In the past, we were not allowed to go out from the house. There was my father-/mother-in-law, brother- and sister-in-law. My husband did not allow me [to go out].
>
> (Older woman Gaibandha)

While there have been significant changes in women's freedom to move beyond the home, there are still significant factors that restrict women's empowerment. Chief amongst these continues to be the elusive concept of honour and the need to protect the family's reputation by continuing to restrict women's freedom by marrying girls at a young age. Critical, too, is the perception held by men that women are somehow less capable:

> The women do not know anything. They are ignorant, they have less brain. We, the men, are wiser than the women … But women's knowledge is (very poor).
>
> (Interview, male, Gaibandha)

One result of this attitude towards women is that they do not have the same access to services, including even such basic facilities as transport:

> Another problem is transport. From 4:30 to 9:00am you cannot get a seat. So many people are in the vehicle. Men have distinct seats but women have not. There is only a box for women. It is not sufficient. So normally we will sit but [men expect] we are to leave this [box] for men [when it is crowded]. Those who come from a good background behave well and allow sick women to sit. [So even] those passengers who have to go far and who have a ticket [can't sit]. We [asked them] to fix some seats for us. They said that no seats will be reserved for the women. Some teachers also [lobbied] on our behalf but got no result.
>
> (Middle-aged woman, Barguna)

Eve-teasing and gossip

The lack of equality for women has several consequences, including its impact on women's empowerment. One particularly insidious form of gender control is the practice of Eve-teasing. In an Australian context this is referred to as sexual harassment. Eve-teasing comes in many forms from verbal abuse, to rape and murder. There are also increasing instances of young women being photographed on mobile phones and photo-shopped into compromising photographs with young men. This assault on their 'honour' is almost impossible to deny or challenge with parents who do not necessarily understand the technology of photo-shopping – nonetheless it has significant consequences for the young women and their families. Girls report being powerless to stop the Eve-teasing and do not know how, or to whom, to complain. This practice can be so severe that it may result in some girls dropping out of school. Yet, as one young girl asks, 'How can the nation march forward if the women lag behind?' Young women discussed Eve-teasing in focus groups:

> I sometimes face problems. Sometimes, the bus conductors tease us. They say this and that. [The girls in the focus group remained silent for some time, not ready to share the problems they experience on way to their schools. They continued]:

> Yes, we do face problems on our way to and from the school. But, boys from our school do not do that. Boys from other schools they say many things; sometimes they take our snapshots with their mobile phones. When the boys try to take photos, the girls lower their heads and leave the place, but they never protest. They do not protest, they are afraid of protesting. If I protest today, I may land in further trouble. Say, they can touch my hand. The girls have to tolerate what they say, but [it is awful]. Yes, some girls inform their parents. The parents advise the boys not to resort to such behaviours but the boys do not pay heed to these requests. If such issues are raised to the local elders, the situation remains calm for one or two days, then the boys again resort to such behaviours. Awareness should be created among the boys about this. They should be reminded that if your own sisters are unable to go to school due to these, then how can other girls go to the school? How can they progress in life? To develop the nation, both men and women should be educated equally. How the nation can march forward, if the women lag behind?
>
> (Young women, focus discussion group Barguna)

An extreme form of Eve-teasing is the kidnapping of girls who are then raped and 'dishonoured' and therefore forced to marry the perpetrator. This practice is more common than may be imagined because it is hushed up by the families:

> She [another young woman] was taken from school. We have one school here. Boys kidnap girls from there. We are women. We don't go out. We just hear it. Men come home, they tell us and we hear it from them. Girls are just taken. Then they are shared. Then they [must] get married. Become wife.
>
> (Young woman, Barguna)

Gossip is another form of gender control often used to impugn the character of young women and to dishonour them. This is providing significant impetus to the practice of early marriage. As one older man described it:

> [When I was young] we got into the water and plucked water lilies; while we were young of course. At that time the girls around our age took a bath there [as well], and plucked water lilies. But then no one ever gossiped about it. But now, if a girl even talks with a boy for just two days in a row; people will start saying that they have some illicit affair or something. The minds of the people are getting so dirty that it beggars description. Girls can't even reach class 9 or class 10, before they are already married off. In our locality the girls are being entered into the wedlock at a very early age. If the children go to school and college won't the boys and girls talk with each other? But even if two classmates talk

> for a few days, the people start to smell a rat. Now you say, what would you rather do? Lose your reputation or get your daughter to marry?
>
> (Older man Gaibandha)

While many are married young, others choose to take their daughters away to reduce the threat of gossip and the threat to honour:

> Yes, security issues are also a major concern. You cannot rear a girl in the village at all. Families with young girls and with financial ability are leaving the village. They don't want to live in the village anymore. They become desperate to leave the village when their daughter is young. Social environment is the main thing.
>
> (Focus discussion group teachers, Barguna)

Early/child marriage

Because of Eve-teasing and the threat of violence and dishonour, there is increasing pressure on families to marry their daughters at a very young age. Respondents in our research rarely spoke about a girl's reputation in a positive sense. There is a basic assumption that girls are not of good character, will have an extravagant sexual life if allowed or are barren:

> If your daughter is not married off they assume that she has some kind of problem and they talk about it behind your back. People can make a lot of mistakes when they are young. But if anything like this happens by chance, you are in deep trouble. The people will start making any sort of gossip centring on that mistake. That is also another reason to have the girls married off early. The father of the girls will naturally want to have their daughter carted off before rumours start to spread. That is the basic reason why we can't wait for our daughter's marriage. Regardless of economic condition, we are compelled to marry our daughters off at a very early age. And people even sell their lands to pull off such a feat.
>
> (Older man Gaibandha)

There are some female village leaders who try to prevent early marriages. In particular the teachers will attempt to persuade families to resist the pressure to marry their daughters early:

> I heard from someone that a child marriage is going to take place in an area. Then, I went there and spoke to the family against child marriage. Then the marriage did not take place .... The girl was not of marriageable age. She was only 12 or 14 years old. She was about to be married off at that tender age. Then we went there in a group and convinced her parents that marrying off their daughter at an early age will damage her health and will create a number of problems – the boy will demand dowry

> later, he will then beat the girl for dowry. The girl's parents understood that.
>
> (Young female teacher Barguna)

Girls married early are often mistreated, expected to act as servants to their husband's family and beaten. Some girls resist and return to their parents' home. The fortunate ones are allowed to stay:

> But my daughter says, 'No.' she says, 'Death is a better option than being there in that [husband's] house again. I won't go there. I won't go to live there. I'll rather face whatever is written in my fate.'
>
> (Older woman, Gaibandha)

Dowry

As a result of the mounting pressure to control girls' behaviour and marry them early, the practice of dowry has been increasing even though it is illegal. However a significant change in this process is that the girl's family now pays dowry to the bridegroom. Previously the groom paid dowry to the girl's family for the privilege of marrying her. This major change in assessing gender and power in personal relations has resulted from women being more visible in public and having education:

> When I got married, there was no thought of the groom getting a single dime. Rather the groom had to pay 100tk to bring the wife along with him. And that money had to be paid to the father of the bride. The groom had to pay the cost of the ceremony. Now, it's the other way around. If you go to any boy or the boy's father, they ask for money or things.
>
> (Older man Gaibandha)

> In the past the women used to cover their whole body and never go out from home. That's why they were attractive to men and they give dowry to marry them but now women are not using the religious code of dress. Men can easily see them. That's why the system has been changed.
>
> (Older man in Barguna)

One potential bridegroom wanted to look over the cow he was to receive as part of the dowry before he met his future wife. This caused the father to describe with disgust his experience with paying dowry:

> Now it is quite impossible to marry without dowry. The amount is 10,000. 20,000, 30.000 thousand and more … now it is quite impossible. Most are not married without dowry. There is a great demand [on the family for] money, though they forget to know the name of the girl but

> not forget about dowry. It is very funny, the cow that is given as dowry, firstly they wanted to see it. Their curiosity is about how nice the cow is. It is like a market of dowry, they take (marry) a little girl but do not take her without dowry.
>
> (Older man Gaibandha)

When asked to explain the change in dowry payment from being an obligation of the groom's family to now being that of the bride's, one man explained the fundamental reason for the change in dowry payments:

> See, the man who has a daughter, he has to marry her off. Or else you have to feed her.
>
> (Interview, older male, Gaibandha)

As noted in Chapter 2, for many women, the family's inability to fulfil the promise of dowry leads to violence.

Summary

This chapter provides a glimpse into the daily lives of rural Bangladeshis, albeit lives that are changing rapidly in line with the emerging market economy and evolving livelihood strategies. Nonetheless they reveal a highly gendered society where women are controlled through social customs and practices. Men dominate the ownership of assets and decision-making and are therefore able to dictate and control the lives and responsibilities of women. What is significant about the evolving nature of livelihood strategies are the changes from a traditional, insular society to one where paid labour is essential for families to survive. This has led to major changes in traditional practices, with men and young girls and boys leaving to pursue income-generating work elsewhere. The movement of married women into the paid workforce has also led to significant changes in traditional practices as women abandon *purdah*, at the same time as they strive to protect their honour. Married women work close to home – their wages and conditions insecure at best.

The persistence of poverty that has led to this drive for alternative income has also resulted in most families taking loans and many women seeking microcredit, often to pay for basic necessities such as food. Food insecurity is a particular issue for all families and is a driver in the searching for alternative income. Women compensate for this by eating less and by attempting to protect their families through income generation.

Gender inequalities remain and are evidenced by a tightening of restrictions on young women and an inflated concern for family honour that centres on unmarried daughters. Eve-teasing, gossip, early marriage and dowry are some of the complications of this significant focus, a focus that many young women find deeply distressing, not least because it threatens their education.

The lives of rural Bangladeshis are difficult, shaped by poverty and food insecurity, compromised by loans and financial hardship and bound by traditional expectations of women and men. Yet there were many times that women and men laughed together while explaining their situations in focus groups – within these tight groupings and communal sharing there is much that strengthens the human spirit. How do these fragile communities react to the threat of significant climate change?

5 Perceptions of climate changes and events, preparations and responses

Rural areas across the world are changing and adapting to major globalizing forces and a pervasive transnational neoliberal ideology that constructs continual improvement in productive capacity as a pre-eminent goal – a goal that places significant pressure on fragile ecosystems and food production systems. The growth of corporatization and the dominance of transnational companies operating above national scrutiny further reduces the power of nation states to regulate and protect the environment (Connell 2011).

Further, the global population is predicted to reach 9 billion by 2040, creating increasing pressures on food production and water security in already destabilized rural areas. The United Nations Secretary General's High Level Panel on Global Security (2012: 13) predicts that this population will require 50 per cent more food, 45 per cent more energy and 30 per cent more water by 2030. Under current global arrangements this may be very difficult to achieve.

This has created a growing conundrum for farmers across the world and for governments seeking to ensure national food security. The emergence of climate change as a new threat to rural areas, with its perverse consequences for weather patterns and production cycles, further destabilizes the lives of rural food producers. As livelihoods become more precarious, commercial production of food has escalated in many countries such as Bangladesh, and a variety of crops are now grown for the market. Often these require loans that lead a vicious cycle of debt, uncertain production conditions and the need to secure income elsewhere to meet repayments.

Export-oriented trade has undermined production for the local market; increased the need for waged work; and accelerated the out-migration of family members to provide income support. There is a growing, mobile, intra- and inter-country workforce exiting rural areas seeking waged work to supplement income. Remittance income earned by a family member and sent back to families has become a common survival strategy for rural and farming families across both the developed and developing world. Bangladesh is no exception. Out-migration is a common occurrence for men and boys going to cities for labouring work and young women moving to the city to work in garment factories or as domestic workers. While the need to source income forces many to leave their rural villages for regional centres and capital cities, the attraction of a secure income, ongoing employment and greater access to services is compelling. This trend results in growing urbanization and, in countries such as Bangladesh, the growth of urban slums where poor sanitation and inadequate services create problems for these expanding metropolises.

A growing wealth divide has emerged across the world, with the richest 5 per cent of people receiving one-third of global wealth (Bauman 2011) and 2.8 billion of the earth's 7 billion people living on less than US $2 a day (UN 2005). While the proportion of people who live in poverty has shrunk, Giddens (2009) notes that 1 billion people in 60 different countries live in abject poverty. In rural villages across Bangladesh, people are experiencing poverty, seasonal hunger, reduced access to services and isolation.

In this chapter we discuss the climate changes being experienced by the people who took part in this research and note the perceptions of climate change amongst this group. We look at differences between the regions and note the development of early warning systems by the Bangladesh government and NGOs. We ask how do the people in the rural areas of Bangladesh perceive the changes in climate that are occurring in their country? Do they corroborate official information regarding climate variability? The participants in our research were asked to describe their perceptions of these changes, their understanding of any variability in the weather patterns and the impacts of these on their lands, livelihoods and well-being. Key informants in Dhaka and background material had given us a critical awareness of the changes, but the way these are perceived at local levels by those who must deal with the outcomes is one of the critical rationales for this research.

There are major changes underway that are impacting on livelihoods, on the types of crops being grown, on food security and on the adoption of various coping and adaptation strategies. As outlined in Chapter 1, there are major changes occurring in weather patterns in Bangladesh. The Appendix outlines the policy responses to these climate challenges in Bangladesh.

As summarized by Khatun and Islam (2010: 11), the predicted future scenario for Bangladesh is as follows:

- Average temperature increases are predicted to be 1.3°C in 2030 and 2.6°C for the year 2070.
- Seasonal variations in temperature are predicted to be 1.4°C in the winter and 0.7°C in the monsoon in 2030, and in 2070 the figure will be 2.1°C and 1.7°C, respectively.
- By 2030, an additional 14.3 per cent of the country will become extremely vulnerable to floods, while existing flood-prone areas will face increasing flooding.
- A 1-metre rise in sea level will inundate 18 per cent of landmass and directly affect 11 per cent of the population.
- Monsoon rainfall may increase by 11 per cent and 27 per cent by the year 2030 and 2070, respectively.

- Up to 40 per cent of investment in development and infrastructure will be affected due to climate change.
- General cyclonic activity in the Bay of Bengal will become more frequent, creating rougher seas.
- There is an increased danger of epidemic morbidity and mortality (mainly due to diarrhoea).
- Increased coastal temperature will cause diseases in the coastal districts.

A qualitative summary provided by a key informant gives some sense of the scale of these changes being already experienced at local levels:

> Salinity is encroaching the lands … affecting the crop cycle and crop production … the weather is becoming warmer, the pattern is changing and … crops that did not flower before September or October now flower almost one month earlier. … when we consulted the farmers, one thing they said consistently, was that their forefathers' knowledge from whom they learned agriculture is no longer of much use, because we really don't know now when to cultivate or not to cultivate. And would you believe that we have fog during summer? Nobody experienced that before.
>
> Then last year, [across] about 10,000 hectares, rice flowering took place before the plants were fully matured because the temperature increased slightly and that affects the plants … they had to mow it down because there would be no grain …. This is already happening.
>
> One of the factors in the decline in the cultivation of wheat in the northern areas is climate change. Because [it is not raining in] winter, and therefore wheat cultivation has been declining in the northern part of the country [where it used to be] increasing. So, there have been a lot of changes – the country is really in a transition in terms of the impacts [of climate change].
>
> (Key Informant NGO Dhaka)

Perceptions of climate variability

How are these major challenges being perceived by those most affected? The perceptions of these changes as expressed by people at small village level in one of the most vulnerable nations on the planet illuminates the reality of living with climate variability. Many took the time to outline their understanding of these changes. These explanations vary from a scientific analysis to a digest of the commercial realities of climate changes to the fatalistic view that it is the will of Allah. In representing the views of village people, we intersperse findings from our survey with comments made by participants in focus groups and interviews.

Participants in each region were asked to give their *perceptions* of weather changes. These questions do not refer to established facts or detailed scientific information. Rather, it exposes the experiences of people in these selected regions. There has been no attempt to influence responses or to seek to modify or enhance claims.

Of significance is that the vast majority of participants note that there are observable changes. In Gaibandha people suggest that winters are more intense and the frequency of floods and river erosion has increased. In Satkhira, which is situated near the Bay of Bengal, people note that the intensity of summer has increased, and the encroachment of salinity from tidal surges in the Bay has inundated once-productive land. In Barguna, also in the Bay region, many people feel that tidal surges and the frequency and intensity of cyclones have increased.

Nonetheless there are significant gender differences in these perceptions, dependent no doubt on gendered livelihood strategies, land ownership and involvement in agriculture, decision-making, child care responsibilities, food security and health issues.

Slow-onset weather changes

Men are much more likely to name and discuss seasonal changes because of their involvement in farming and other land-use activities. Women are much more likely to comment on temperature or rainfall change because of the impacts on household activities. Women note the increase in water and airborne diseases and the threat of these to health, to agricultural activities such as seed saving and to household food production. Despite these gender differences in perceptions, respondents were very definite about weather change. Only 19 of the 617 survey respondents answered that they did not know if there had been weather changes. Those who participated in focus groups and interviews were also quite definite about changes:

> Seasons have changed. It is no more as it was before … now it has become disarranged or disordered.
>
> (Older man, Barguna)

Amongst survey respondents, changes noted related particularly to seasons, temperature and rainfall, with some wind changes and increased cloud cover cited particularly in Barguna.

Seasonal changes

Respondents were asked the reasons for the seasonal changes that they perceived. The following is a summary of the top three reasons given for each *perceived* change.

Box 5.1 Top three responses for types of perceived changes in seasons

1 Shorter rainy season (33 per cent of respondents)
2 Change in the number of seasons (30 per cent, and more likely in the south)
3 Unpredictability of weather (20 per cent)

Seasonal changes were noted particularly in relation to agricultural production. Traditional practices are based on localized understandings passed through generational transfer of knowledge. Perceived weather changes have destabilized this knowledge and indicate marked impacts on traditional production practices.

> All the things like weather, seasons, winter, summer are totally changed. Now the cold is not as severe as before. But the summer exists for longer periods. Rains are irregular which is bad for cultivation. And now salt has increased in the field which is bad for regular crops.
>
> (Male focus group, Barguna)

> Another problem is the change in the climate over the last 20 years. Previously it was feasible to cultivate as there were six seasons, it would rain when it is needed. But now, nothing is happening as it should be. We do not get rain when needed, or it is heavy rain causing damage to the crops.
>
> (Male focus group, Satkhira)

Table 5.1 Respondents who answered, YES there was change

Climate factor	*Women*	*Men*	*Total*
Seasons	38%	66%	53%
53% of all respondents reported season change 60% Barguna, 55% Satkhira, 43% Gaibandha			
Temperature Change	85%	93%	89%
89% of all respondents reported temperature change 94.4% Satkhira, 93.8% Barguna, 78.6% Gaibandha Women are much more likely to report temperature change than any other weather/climate variable			
Rainfall	78%	82%	80%
80% of all respondents reported rainfall change 91% Barguna, 77.6% Gaibandha, 71% Satkhira			
Wind	54%	58%	56%
56% of all respondents reported wind change 70% Barguna, 55% Satkhira, 42% Gaibandha			
Cloud cover	30.5%	45%	38%
Only 9 respondents said they don't know and only 19 said they couldn't remember any changes			

This has led to changes in production practices and even changes in the types of agricultural products grown. This is particularly evident in Satkhira where there has been a major shift from crop production to shrimp-farming on land that has been inundated by saline water. This has also led to conflict and co-option of land by what some refer to as 'the shrimp barons'. Traditional communal agricultural practices have been totally destroyed by the co-option of land and the new dominant product. Shrimp farming is also exacerbating saline intrusion into once-productive soils.

> Shrimp farms are benefiting 10 per cent of the people while the remaining 90 per cent are financially losing out. Now rich people are becoming richer, poor are getting poorer. But previously it was not like that; people with little land would also have a decent living through a little paddy cultivation. So the cost of paddy cultivation has increased due to climate change.
>
> (Male focus group, Satkhira)

Seasonal changes have significantly affected trees and groundcover, noted by many women, particularly in Satkhira. They also suggest that previous practices that relied on the short-term inundation of land for the irrigation of crops has now been destabilized as the waters do not retreat but stay on the ground creating significant salinity problems. For women this has had a marked impact on their access to fresh water for the household and on their small-area household food production. They must go further to source fresh water, often walking through dirty water to do so. This has had a significant impact on their health, causing skin diseases and other health problems.

> In the past the weather was good. We observed rainfall during the rainy season, warmth in summer and cold in winter. Now it is not. There are big changes. In the past there was sweet water everywhere, there were trees and we could cultivate paddy and vegetables. But now the weather is different. There are no trees; we are water logged, because of saline water the trees died. There were many big trees here, we could not see the other house for trees, and people could not see our house from the road. There [used to be] lots of trees here.
>
> (Older woman, Satkhira)

Temperature changes

Box 5.2 Top three responses for types of perceived changes in temperature

1 It is hotter during hot months
(72 per cent respondents: 72 per cent women and 71 per cent men; 77 per cent in Barguna, 76 per cent in Satkhira and 62 per cent in Gaibandha)

2 It is colder during cold months
(62 per cent respondents: 65 per cent women and 59 per cent men; these were evenly spread across districts)
3 There is an increase in temperatures overall (46 per cent respondents, including 55 per cent of Barguna and Satkhira residents and 28 per cent of Gaibandha residents).

Respondents were quite clear about increases in temperature and noted definite patterns relating to the temperature being hotter in hot months and colder in cold months. Women are much more likely to report changes in temperature than any other weather changes:

> [In the past] summer was not extreme like now. There are no trees, only sands like desert. There are only sands. Yes, the winter does not come at the right time, … we do not feel the summer in time. There is no regular flood. In the past the winter started from the middle of the Bengali month of Ashwin. There was cold along with rain and wind. But now that is not happening … We are tired of heat, extreme heat.
>
> (Older woman, Gaibandha)

> Nature and the weather have been changed totally. When it should be cold in winter then we can't observe that much cold. In rainy season when we need rain … we don't get any rain or very little rain.
>
> (Young woman Barguna)

Rainfall changes

Box 5.3 Top three responses for types of perceived changes in rainfall

1 There is a decrease in rainfall overall (56 per cent – higher proportion in the south)
2 Rain falling at unexpected times (48 per cent overall)

(Higher proportion in Barguna: 58 per cent)

1 Rainfall is late (25 per cent of respondents).

Only 7 of the 617 respondents report no change.

The effect of uncertain rains on food production is significant for people who rely on growing their own crops for food security and household survival:

> When we need rain we don't get it; for example in the month of Vadro (monsoon) we need water to sow the seeds of paddy. But due to lack of rain we can't grow good crops.
>
> (Older man, Barguna)

> Here the main problem is erratic rainfall. Because of the changing climate, rainfall and flooding do not occur in time. That's why people can't cultivate their land easily or properly. The type of food cultivation we expect [with our usual] weather is not possible for the changing climate. For instance, GANJIA/GAZI rice (one kind of rice) was cultivated in [previous] years but this year it is rare.
>
> (Male focus group Gaibandha)

For some, like those in the Char areas, the lack of, or uncertainty of, rain is compounded by flooding and river erosion:

> We have now less rainfall and the weather is extreme. Land has become deserted. Sometimes due to abnormal floods we also face problems in cultivation.
>
> (Older man, Barguna)

Changing rainfall patterns and the added burden of salinity is a critical factor for women who practice seed-saving. Seeds are being damaged by dry, saline conditions:

> When it is winter season … there is no sign of winter. Rain is needed but [there is] no rain. Rainfall is unusual. Whenever something is needed, it is scarce. Seeds are damaged by the over-excess of rainfall. When rain comes we want to plant seeds but no seed is available. Because of salinity, seeds have been damaged.
>
> (Young woman, Barguna)

Some involved in farming note that the changes have led to the adoption of new technologies such as irrigation:

> But now, it rains some time and some time it does not. Then we have to irrigate. We irrigate by machine. But in any case we do not leave the lands uncultivated.
>
> (Older woman, Gaibandha)

Wind and cloud cover changes

Box 5.4 Top three responses for types of perceived changes in wind

There was less discussion amongst participants of changes in wind or cloud cover and the overall percentage of responses to these variables was not large. Nonetheless stronger winds were reported by 22 per cent of survey respondents of whom half were from Barguna. A further 16 per cent reported less wind overall.

Top three responses for types of perceived changes in cloud cover

Changes in cloud cover were reported by a small percentage of respondents, some suggesting that there is less cloud cover (17 per cent) and some that there is cloud cover at different times (16 per cent). Barguna respondents were more likely to report these changes.

Catastrophic weather events

Respondents were asked to comment on perceived changes in catastrophic weather events as opposed to climate variability. A large majority are quite clear that there have been significant changes in catastrophic events such as cyclones, floods, storm/wave surges, as well as slow-onset events such as river erosion, drought and extreme temperature – and that both catastrophic and slow-onset events have an impact on their lives. Thus it is evident that there is a need to focus on both catastrophic disasters and slow-onset climate events to gain a fuller understanding of the impacts of climate events on the people of Bangladesh. Respondents were more likely to comment on increases in the intensity of cyclones, flooding and water logging and the extent of drought conditions.

Many survey respondents noted particular climate events as having a devastating impact on their lives and most commented on multiple events as indicated in the box below.

Box 5.5 Those who commented on an increase in the number and severity of particular catastrophic events

1 Flood 81 per cent (84 per cent of women and 79 per cent of men)
2 Extreme temperature 69 per cent (70 per cent of women and 68 per cent of men)

Highest in Barguna 164/210 participants (78 per cent)

1 Drought 61 per cent (61 per cent of women and 61 per cent of men)

Highest in Barguna 152/210 participants (72 per cent)

1 Storm/wave surge 59 per cent (68.5 per cent Barguna, 63 per cent Satkhira, 45 per cent Gaibandha)
2 Waterlogging 10 per cent
3 Cyclone only 9 per cent

Floods and storm surges

Participants in interviews and focus groups also explained how these catastrophic events had impacted on their lives. Most noted that the intensity of cyclones and floods differed from previous experiences:

> Actually we saw many floods before Cyclone *Sidr*. Earlier floods happened [at about a] 10- to 15-year interval. Stormy winds, rain, cyclones, etc. are common. But *Sidr* was different. We never faced this type of terrible cyclone.
>
> (Women's focus group, Barguna)

The storms create storm wave surges across many areas. These are particularly pronounced in coastal regions. Because of the increasing frequency and intensity of these storm surges the government and NGOs have been investing in rock walls to attempt to reduce their impacts:

> Storm surges come with a high peak, so we have boulders quite well supporting the boundary in the coastal areas. I mean our parliament is simply protecting people of these areas from storm surges. But because of the changing pattern of the weather, it is happening more frequently, and with greater severity.
>
> (NGO Key informant Dhaka)

The impacts of long-term salt water intrusion and inundation on previously productive land and the consequent ecosystem degradation and impacts on their livelihoods is noted by many of the men who took part in focus groups:

> There were lots of trees when we were younger. The quantity of salty water was less … Lots of sand came after I. This land was a river and now it's filled with soil … Previously there were lots of fish in the river and we used to catch them … Now there are no fish in river … Besides fish are not getting sufficient food in the river.
>
> (Older man, Barguna)

Women across the three regions noted the devastation to their livelihoods caused by storms and flooding. Storms and floods have taken away their crops, flooded and/or destroyed their homes, and corrupted their water supplies, often leading to significant issues with finding fresh water and carrying this back for the family's needs.

> This last year salty water entered our area due to the destruction of the embankment. The quantity of salty water entered was more than the [normal] rain of Asar season (Bangla monsoon). We suffered a lot.
>
> (Older woman, Barguna)

> There are storms. We plough and seed and the flood takes away the crops. Crops are damaged time and again. [We have] storm, floods, drought in our area. The crops die for drought. Suppose we are going to cut the crop, but flood takes it away. This year the jute is damaged by flood. I sowed jute this year. But the river took it away. Now I shall have to owe money to run my family. I do not have a cow or a goat. What can we do?
>
> (Older woman, Gaibandha)

> There has been flash flooding that inundated our households, washed away fish from the farms and ponds. This year the flood came only after 15 days of sowing the seeds, so all the crops were destroyed. Later we started again by irrigating with water from the ponds, but we do not know if we will get a good harvest because the pond water is all saline water.
>
> (Women's focus group Satkhira)

River erosion

One of the critical impacts of major flooding and volatile weather patterns has been the ongoing process of river erosion in many parts of the country:

> So, the river, when it gets wild, it kind of washes away underneath and then the whole bank collapses in … [in one area where we were working] their school and their market all just collapsed into the river within a few hours.
>
> (NGO Key Informant, Dhaka)

Although river erosion is not new, it is arguable that the pace and scale of the changes are more intense. Men in focus groups across all three regions noted discernable changes in their lifetimes in relation to river erosion, the depth of rivers and transport difficulties:

> Before we crossed the river by swimming but now it takes 10 minutes to cross the river by engine boats. This erosion started from our birth period. Our fathers also said the same. I saw four roads besides this river. But now nothing is visible.
>
> (Male focus group Barguna)

> Yes, about 70 years back people used to walk across the river. The current is too strong now.
>
> (Male focus group Satkhira)

> [We are told that] in the ancient time the river was so deep … At present the rivers are being silted and that's why there is a lack of river water. Now rivers are not so deep. River erosion is increasing. We are confused that we will not be able to continue our livelihood in this Char. The present situation is very alarming.
>
> (Male focus group, Gaibandha)

The frequency of climate events

Participants in our research noted the constancy of weather events – that they were always on the edge of a disaster whether it be a slow-onset drought, high temperatures, encroaching salinity, river erosion or catastrophic events such as cyclones or floods. There is a distinct note of resignation and even despair amongst respondents about the unpredictability of weather. This encompassed not only village participants but also Key Informants such as NGO workers:

> We work in some of the most remote and difficult areas and it's had a direct impact, because the frequency of disasters, the intensity of disasters has changed. When I joined [international NGO], I came back from the UK to take up this position in June, 2007. By July, August I had two rounds of floods. November Cyclone *Sidr*. There was an earthquake, there was river erosion, flash floods … I couldn't keep my head above water most of the time. So that's the intensity in just six months – and then came [cyclones] *Aila* and *Nargis*, one after the other.
>
> Whether you want to link it to climate change or not, it's really up to you. But for me, where I'm sitting, it seems why is this happening? The only explanation I can find is that there has been change in terms of the climate behaviour, whether it's a result of development or planning. And when we talk to the people on the ground they say, 'Something is not right. It was not like this. We would not have two cyclones back to back.' And we have seen that. We had the cyclone in 1970, then we didn't have any until 1988, '91, 2007. And then 2007, 2008, 2009, I mean this is ridiculous.
>
> (INGO Key Informant, Dhaka)

People who live in the villages and must deal with these impacts on a daily basis note the same sense of helplessness as the NGO workers. The impact on the devastated landscape is a constant reminder to villagers of the constancy and severity of these events:

> There were lots of trees when we were younger. Animals and trees have now decreased by 25 per cent. The quantity of salty water [has] increased.

> Lots of sand came after *Sidr*. This land was a river and now it is filled with soil. Previously there were lots of fish in the river and we used to catch them. Now there are no fish. The rivers were deeper … but now fish cannot stay as the water level is decreasing.
>
> (Older man, Barguna)

Summary

There is no doubt that the people who took part in this research are aware that there are significant changes in weather patterns in their areas. These changes include seasonal variability, temperature rises, wind changes and cloud cover variations. People also report an increase in the frequency and scale of extreme events. These changes vary across the three regions. Nonetheless there is consistency in the reporting. Men and women are likely to notice different aspects of the slow-onset changes, with women more conscious of temperature changes and men of the seasonal variations.

6 Negotiating the climate disaster experience

By contrast with the slow-onset changes outlined in the previous chapter, this chapter focuses on catastrophic events in order to outline the experiences and actions of people before, during and after major threatening events. The chapter examines the way women and men describe the event and negotiate the immediate, medium- and long-term effects of the disaster. Our research has been particularly focused on slow-onset events – drought, variable rainfall, river erosion, saltwater inundation and rising temperatures and sea levels. This focus has been deliberate because of a lack of critical information on the changes these events cause over time. Nonetheless, in the context of Bangladesh, it is difficult to ignore the impacts of frequent and devastating climate catastrophes – the cyclones and floods that are occurring with alarming frequency. These are the experiences that are seared into the minds and hearts of the people who have been through extreme circumstances.

More recent research undertaken in Bangladesh suggests three stages of a disaster – 1) the preparation for the event, 2) managing the event and 3) the recovery phase. Individuals and families develop coping strategies during preparation for, enduring and recovering from natural disasters (WEDO 2008):

- *Preparation or prevention strategies* include predicting and preparing, teaching children, protecting the household and storing items.
- *Managing or enduring strategies* include ensuring familial safety and food security, protecting assets, managing finance, migrating, seeking employment alternatives, accessing government and NGO agencies for assistance, skipping meals (Khandker 2007), borrowing money, increasing microcredit, making distress sales of assets and adopting natural health options (Kitkiri 2011; Nasreen 2008; Brouwer *et al.* 2007; Khandker 2007; del Ninno and Borsch 2003). In cases of extreme deprivation, some women and girls engage in sex work and sex tourism (APIT 2009; Save the Children 2006).
- *Recovery* strategies include rebuilding houses, securing an income, repaying debts, restoring livelihoods and treating affected family members. People in our study describe a fourth discernible phase of a major

climate event regardless of whether it is a sudden catastrophe or slow-onset event – the long-term recovery stage when adaptive responses (both positive and negative) and new regimes are established.

The four phases described by our participants include preparation for the event, managing the event itself, the immediate recovery period and the long-term process of adaptation. This chapter focuses on the first three, noting the types of preparations taken, people's experiences during an event and the immediate post-crisis reconstruction phase. Long-term adaptations taken by our participants are discussed in Chapter 7.

In our observations of the way people describe their experiences, we note a slow erosion of resilience as people find that the number of events challenges their capacity to re-establish their lives on a regular basis. For example, in one village affected by slow, but constant river erosion, the villagers had moved three times in the last two years as the river had taken massive chunks of their land in a steady erosion of their lands and livelihoods. Each move requires rebuilding their houses and re-establishing their occupations and incomes. Such pressures gradually erode resilience and the capacity to cope with additional changes. People describe such major challenging experiences with a sense of despair:

> Here human lives have no value.
>
> (Women's focus discussion group, Barguna)

> [Cyclone] *Sidr* has broken all the pillars of our life.
>
> (Older man, Barguna)

Before a disaster

There is much that is positive about the way this tiny country has addressed the challenges it faces. The Bangladesh government and NGOs working in the rural regions have invested heavily to build disaster preparedness. Strategies developed include providing early warning systems to alert people to an imminent weather event, constructing shelters and other infrastructure such as rock walls to hold back storm surges and raising awareness through the distribution of information on climate changes.

The early warning project being developed by the government as part of their disaster preparedness has required intense planning and development of processes from the very top of government. Early warning of an approaching catastrophe is critical to how well people can prepare and survive. However, given the diversity of the landscape, the fragility of infrastructure, the isolation of many of the most vulnerable areas, the scale of the disasters and the magnitude of impacts, early warnings are not always possible:

> The Disaster Management Council for National Disaster Management is headed by our Honorable Prime Minister … We can calculate the

> movement [of the cyclone] and when it may hit. Then accordingly we start informing people that it's a storm; it may hit in such and such a place, it is moving at such and such a speed in this direction, there's a possibility it will hit in Bangladesh.
>
> Every five minutes we phone into the interconnect meeting.
>
> At the same time, we have volunteers in coastal areas, – 47,000 in number. And the … volunteers, 30 to 40 per cent are women. These volunteers, they help people to go to shelters, they help inform people about the cyclone. We [contact them by] wireless, and the internet. We inform them where it may hit. Those volunteers, are in the union … [Volunteers] have to take the old people and women to the shelter, have to take children to the shelter, they help to take all to the shelter. They are very dedicated, we don't pay anything to them. They feel it's an honour to be a volunteer.
>
> (Key informant, Dhaka)

NGOs are also crucial to disaster risk reduction, planning, monitoring, reconstruction and rebuilding capacity:

> At [NGO] we are concentrating more on vulnerability assessment and risk reduction – so it's more of a capacity development exercise that [NGO] is now undertaking in the field of disaster management, and that's cross-cutting so we don't differentiate. I think we will be training all the health volunteers. We have about 80,000 of them from each village, all women. So, the whole 80,000 will be taught the basics of how to react pre, during and post disasters. The school teachers, about 50,000 of them, pre-primary and primary, will also be given an orientation. …
>
> Our role is preparing communities and [NGO] staff in reacting or being prepared. And then also inculcating an attitude so that they can reduce risks, so that they aren't stupid enough to do things that will exacerbate this sort of thing. This is one. Second is so there is a uniform approach to a disaster. The frequency and the severity of these events will probably increase over time, so if they increase, we have to get prepared for it.
>
> (NGO Key informant, Dhaka)

Early warnings

Sometimes, however, early warnings are not received, and often those who are at home may not receive them. Access to information is critical to adequate preparation and responses but is fraught in a country where there are limited points of contact. Of concern too is that when people receive a warning they often cannot imagine the scale of the impending disaster and may in fact ignore the warnings. Our participants were asked about how they receive early warnings and what is likely to happen in the immediate preparation phase. What is evident from these responses is that the receipt of early warnings is not guaranteed and that even when they are received they may not be taken seriously because people do not consider that the scale of the event will be different to previous storms or cyclones.

Demonstrating the difficulties associated with information dissemination, men in Barguna noted that they heard about *Sidr* via a BBC radio programme broadcast from the neighbouring country of Myanmar:

> We heard the BBC and we got the news from one of our relatives from Myanmar. The signalling [early warning] system of Myanmar is more developed than that of Bangladesh. He got news from that channel and immediately informed us.
>
> (Male focus group Barguna)

Others noted that the speed with which an event occurs means that early warnings may not be received and many, particularly women and children, may die:

> Most of the time we get a signal [early warning]. But we don't get a signal if the storm starts suddenly. … Women and children along with the elderly and cattle go to the shelter. Some start crying and lamenting for their husband and children [lost in the water] … we do not have life-jackets. If we had, the death rate [from *Sidr*] would have been reduced.
>
> (Older male Barguna)

Others report not taking the warning seriously because they thought that the scale would be no different from previous events and their embankments would hold. This occurred in both Barguna and Satkhira. In the latter, men cut the embankments thinking that would release water, in fact it only aided the inundation and flooding:

> No, we didn't get any warning. Even if we had got it, we didn't care [take it seriously]. This is the important thing. We thought that the water wouldn't come crossing the embankments. So we didn't care. This is the main point.
>
> (Older man Barguna)

> At first there was a bit of rain that caused waterlogging. So we started repairing the embankment. At one stage we cut a portion of the embankment to get the water out. In the meantime a person called and informed us that their area has been hit by [cyclone] *Aila* and suggested to us that we move to shelter quickly. Before that none could believe that [village name] would go under water too. We did not receive any early warning from the government or NGOs.
>
> (Satkhira male focus discussion group)

The sheer scale and force is terrifying and, warnings or no warnings, people are ill-prepared:

> When the *Sidr* started I was at my home. At that time without any warning the thunder started. All the men and women took shelter in the building beside us. At around 12 pm the tidal surge started. The tide was like a tornado with a strong current. We had no prior warning about it.
>
> (Older male Barguna)

Women and early warnings

Women are far less likely than men to receive early warnings because they may not have access to mobile phones, radios or televisions and they may be isolated in their homes and are restricted in their movements by customary practices. Previous work suggests that there are significant challenges for women in coping with climate events resulting from:

- limited access to early warning information, critical services and facilities, financial security, markets and communication, and decision-making arenas;
- the social expectations of the *appropriateness* of women's actions in response to warnings;
- increased responsibility to the household;
- difficulty in accessing relief goods; and
- physical constraints (such as long hair and clothing).

(WEDO 2008: 57)

This research is corroborated by others who note that women are often not informed about an impending natural disaster or about subsequent relief support (Oxfam GB and Naripad 2011; Ahmed *et al.* 2008; Nasreen 2008).

Because women are dependent on men to share early warning messages and have much more limited access to information, they may not receive warnings, or men may not have the time or inclination to pass them on. Customary law suggests that women must not leave their compound without a male escort or at least without permission. In the previous chapter we noted that most women do not make any decisions about their livelihoods or actions, therefore in a crisis they may be immobilized by a lack of direction from their husband or male relative. As a result many women perish because they do not receive information or do not have permission to act on the advice. Women and girls also report feeling reluctant to go to the shelters when there is no privacy or safety and, therefore, where their honour may be impugned.

Safety is a critical concern for women during and after a disaster – as their reputation and the honour of their family is dependent on their actions. However, even when informed, women's mobility (including seeking refuge in cyclone shelters) is significantly restricted by practices such as *purdah*, prohibited inter-gender interaction, the risk of sexual abuse when walking long distances, social condemnation of women swimming, running or climbing trees, and a lack of transport (Oxfam GB and Naripad 2011; Ahmed *et al.* 2008; Ahmed 2008; Nasreen 2008; Cannon 2002; Masood and Wahra 1997). Furthermore, women's clothing restricts mobility during and after a disaster (Ahmed *et al.* 2008; BRIDGE 2008; Cannon 2002).

Many women die waiting for relatives to return home to accompany them to safety (WEDO 2008). They may also be reluctant to leave because evacuation may mean the loss of cooking utensils, goods and livestock required to ensure family food security (Oxfam GB and Naripad 2011; Nasreen 2008). Nasreen (2008) summarizes the reasons women and girls may stay in the home rather than seek refuge as: a lack of shelters, fear of losing houses and essentials, fear of security for young girls and young women, fear of losing honour, lack of cooking and toilet facilities at the shelter and difficulties breastfeeding in front of strangers.

When women do access cyclone shelters, they may be compromised by the crowded conditions, lack of privacy, safety and sanitary facilities, limited sleeping spaces, and no designated space for lactating mothers and pregnant women (Kitkiri 2011; Ahmed *et al.* 2008; WEDO 2008).

Young women and girls are particularly vulnerable during a catastrophic event. They may be unable to attend school (Nasreen 2008) and in shelters they are humiliated and endangered by the lack of space and privacy (Save the Children 2006). Verbal, physical and sexual abuse of children and particularly girls occurs, as well as abduction and trafficking. Girls experience challenges in daily tasks such as bathing, using the toilet and sleeping in exposed living conditions with unknown men. Rape and harassment of young women is reported (Ahmed *et al.* 2007). Homelessness, skin lesions (due to waterlogging) and impurity through sexual abuse further compromise young girls (Ahmed *et al.* 2008; Ahmed *et al.* 2007; Rashid and Michaud 2000). Their ability to maintain their reputation and honour in such circumstances is compromised (Rashid and Michaud 2000) and early marriage of girls increases as a result (Save the Children 2006). There is some suggestion that opportunistic men may marry vulnerable young girls and then sell them to brothels (Ahmed *et al.* 2007).

Because of the significance of this previous research in assessing the gendered responses to catastrophic events, we asked our participants to comment. Our research corroborates previous findings – women may not receive early warnings and, in some areas, men went to the shelters without informing their wives of the danger:

> Women and children didn't [necessarily] get any early warning messages. This is a great problem. So they are staying at their home and at that time most of the men went to the shelter or out of the area … Our early warning systems are … gender biased.
>
> (NGO Key Informant, Dhaka)

When women do receive a warning, they may wait for instructions from their husbands before acting, thus endangering their lives. If men ignore the warnings, women are further endangered:

> It has been observed that, if the women try to take any step before a disaster, the men would stop them. Suppose if we say that there has been a cyclone warning, the men ignore that. The problem is they do not value us unless they are in trouble. If we want to move to safety before a disaster, they would not allow us. They even stop us from taking preparedness measures such as preserving dry food. If we ask them to strengthen the house before the season of storm, they would not listen to us. Even many women are not allowed to touch the TV or radio. So they have little chance to get the early warning.
>
> (Older woman, Satkhira)

Women are very much dependent on men for permission to move beyond their compound and this has significant repercussions during a disaster if they must wait for this approval and a male escort. Further, many report that they may not receive early warnings:

> Some of the well-known cyclone shelters in the coastal zone of Bangladesh are not properly being used by females as much as by their male counterparts. That's because there are gender issues – when to leave the household for example. And the females – when they go there they might have to stand in the presence of hundreds of other males for maybe 36 hours at a stretch. But then the toilets are at the ground level … nobody would dare to come to ground level. The males can urinate in the presence of many others, but, for the Bangladeshi women, it's out of the question. Then of course they feel shy, they know the social repercussion of doing that obvious thing and then eventually they refrain from going out of their home and assume the risk of being drowned.
>
> (NGO Key informant, Dhaka)

Many women reinforce that even when there is shelter, they choose to stay at home:

> I said that if I die then I would prefer to die at home.
>
> (Young woman, Barguna)

There is also some suggestion that child abuse occurs in shelters adding to the reluctance of women to go to shelters with their children:

> I mean in Bangladesh, child abuse is not a topic that you normally talk about, but slowly I think the taboo has been broken. So, [NGO] is trying to stay a step ahead and trying to understand, and then building up a

> body of counsellors, who can be located across the country as first responders. So, the people who we are targeting to be trained are basically going to be the identifiers, if you want to call it that way.
>
> (NGO Key informant, Dhaka)

Preparation

Participants were asked to comment on the ways they prepare for catastrophic weather events – what are the significant tasks that must be addressed? How do they prepare their families, their homes, their possessions and their livestock? There has been considerable research done on the way people prepare, manage and are impacted by disasters. Participants in our research note that there is often little warning of a major event, but that when there is they swing into action. As one woman in Gaibandha explained, her preparations when a disaster is imminent are extensive:

> I need to collect firewood quickly, bring the cattle to a safe place, and arrange everything properly.
>
> (Older woman, Gaibandha)

People note that if they have enough early warning that a disaster is coming, they prepare by storing food, preparing the house, raising goods into the roof and trying to secure livestock. There is a distinct gender division of labour involved in this phase, with men being responsible for tasks outside and women with household tasks:

> We store rice before any disaster. We make rice from paddy by machine.
>
> (Young woman, Gaibandha)

> [Men] try to save food as well, buy things we might need during an emergency. But usually, they tackle the outside business, the women manage the internal household.
>
> (Satkhira women focus discussion group)

The lack of time to prepare is a factor in survival:

> [People do not have much time to prepare] they look for their children, their boat, their cattle. They put them in a safe place or on the rooftop or somewhere and then, most of the time they can't reach the shelter houses … To survive they should go to there. And that's the only way … those who survive do not survive by themselves, they survive along with their children and along with their resources, and even a small amount of food; you would be surprised how they manage the risk, the extreme kind of risk. This management capacity, this is the urgency on the part of women.
>
> (Workshop Dhaka)

Women's responsibility for their households and families continues in a natural disaster, with expectations that they will protect property, shepherd children and livestock to safety, and see to the family's survival, including building makeshift shelters and planning food security (Masood and Wahra 1997):

> when a hazard strikes she has to safeguard all the belongings and any hint of an asset, send everybody to safer refuge, and stay back to ensure that the household is not ransacked while in absence. In this process, she accepts whatever the consequence of the imminent hazardous event. In a quickly on-setting hazard event, such an apparent 'passive' response may lead to death and such a risk is deliberately undertaken by women.
>
> (Ahmed *et al.* 2008: 5)

Following a disaster, this normative household management role expands to include queuing for relief, fetching safe water, reconstructing housing and providing psychological support to surviving family members (Ahmed *et al.* 2008). Women's tasks are made more difficult by the loss of utensils and other household essentials, food insecurity, flooded kitchens and considerable difficulties associated with fetching safe water. This may involve walking as much as 10 kilometres and taking up to 4 hours, irrespective of their own physical condition (Ahmed *et al.* 2008; Nasreen 2008; WEDO 2008; Cannon 2002). Further, waterlogging causes women to become entrapped in their homes, limiting access to safe drinking water and sanitary latrines, and sometimes forcing them to shelter on rooftops (Nasreen 2008; Ahmed *et al.* 2007).

Scale of damage

The scale of the damage from more recent catastrophic events is something for which various levels of government, NGOs and most people are unprepared. In Satkhira, for example, one public official reported there was not one shelter for the 9,000 people in his ward. He recounted that, in one village during a catastrophic cyclone, people tried to reach a banyan tree that provided the only hope of survival. A young woman from this village related what happened next:

> We didn't understand that the village would be flooded this much! Before even we could understand, the water came. When I looked inside the room [my home], the water [was] inside. The water was up to my waist … There was no shelter in our locality. Here in [village name] there was a banyan tree. The men, the boys took shelter. Some of them could not swim across to the other side (near the banyan tree). People were lost! So many people died! Those who lived, they lived through their strength. They held on to trees, they floated.
>
> (Young woman, Barguna)

When a catastrophic disaster hits and people are unprepared they may be swept away by floodwaters. One young woman in Barguna explains how her clothing held her back – to save her life she took off her sari. Given the dominance and prioritizing of women's 'honour', this was a major decision:

> The sari gets whirled around the body. Men wear lungi, so they can put off the lungi and swim to save [themselves]. Women have a thing of honour – that is why women suffer more, men less so. Men can swim or even climb a tree, but where can the mother go leaving their children?
>
> (Young woman, Barguna)

Loss of possessions

A young woman from Gaibandha explains how a sudden-onset event swept away her house and how vulnerable this made her feel when there was nowhere to go. One of her immediate concerns was that her children not fall into the water. In the immediate aftermath she must clean up the mess, reconstruct her house and try to protect her family's health:

> After dinner when we went to bed, there was a possibility that the river may take away my house. Water comes, we raise the upper bed mattress, bed sheets and other things so that they cannot get wet. We lift the bed and boxes. We ensure that the children do not fall into the water. We have to be careful about that. After the water lessens, our wooden furniture and the pillars of my cottage break after the flood. Soil makes the inside dirty and I need to clean them after a flood. I need to reconstruct my cottage. I need to clean everything. Someone in my family will be affected from diarrhoea or cholera, it makes the clothes dirty. I need to clean them. That is what we have to do [after a flood].
>
> (Young woman Gaibandha)

An older man from Barguna explains the despair that follows an event that destroys the hard work of a lifetime – the cyclones, he explains take not only his family's possessions but also their happiness:

> I built my house at a cost of 70,000 taka. I built that house with lot of difficulties. I had no land. I used to work hard on other peoples' land and started saving money. I managed to save that way for years and finally collected the money and built that house. Then I bought a cow and started selling the milk of that cow. The milk was pure. We do not contaminate milk with water or anything else to make it [last longer]. We were happy. And then *Sidr* came; and then *Aila*. They snatched all our belongings and our happiness.
>
> (Older man, Barguna)

Several men echoed this feeling of despair. In focus groups and individual interviews, men explain the utter devastation associated with losing their livelihoods, possessions and well-being. Men in a focus group in Barguna described the difficulties associated with feeling so desperate and yet needing to keep on going, despite their losses:

> *Sidr* killed us. If we died during that time, we would be relieved. Now we are facing various problems. *Sidr* has just thrown us into piles of problems. But we have to exist.
>
> (Male focus group, Barguna)

Women's livelihoods are significantly affected by natural disasters and are difficult to restore, with the loss of livestock, harvest and homestead, difficulties fetching water, shortages of food and other necessities and inflated prices, limited access to markets and loss of income, savings and employment (APIT 2009; WEDO 2008; Cannon 2002). Women thus experience greater economic insecurity. Further, waterlogging, which can last for several months, reduces local employment opportunities (Ahmed *et al.* 2007). Their increased workloads limit their capacity to participate in other, non-traditional activities (see Neerormi 2010) and a lack of land ownership impacts women's access to housing and shelter (Nasreen 2008).

Immediate reconstruction phase

During the immediate aftermath of a major event, people are heavily reliant on government and NGO support for the very basics of their lives. Food, materials to build temporary shelters and other survival tools are supplied in the early days:

> They gave rice, lentil in a bag. This is how we manage. One NGO gave 18kg of flour. And they gave a VGF card [social security card] to get 15 kg rice every month. We collected vegetables and with those we just survived with the children. Then my husband started working.
>
> (Young woman, Barguna)

A young woman from Satkhira describes how she constructed a shelter from tin, bamboo and wire supplied by an NGO:

> [During that post-disaster phase] the NGOs came here and collected data. They noted how many cottages had been damaged in this ward, how many cattle were destroyed – they took all the data from here. After taking the data they helped me with 10 tins, 10 bamboos, some wires and some pins because I did not have anything at that time [to build a house]. After that I made a cottage with that and other things and I arranged my house.
>
> (Young woman, Satkhira)

Access to clean drinking water is one of the highest priorities in the immediate post-disaster phase. People may be surrounded by contaminated water and waterlogged ground, but often they have very precarious access to fresh water for drinking. This has a significant impact on health and household tasks. Collecting water is a traditional female task that may become overwhelming in a post-disaster situation:

> My prayer [ritual], cooking and bathing all are dependent on water from outside [the house]. This is so painful – we cannot tell some things to our men. Even they do not think about our pain and they do not try to help us. But after getting training from an NGO the situation is changing and they are trying to help us.
>
> (Young woman, Barguna)

The reliance on outside help in the immediate post-disaster phase leaves people highly vulnerable to food and water insecurity, a lack of shelter and mounting health issues. People depend heavily on support coming in as quickly as possible. There was some frustration expressed at the lack of response and aid received in some instances, despite publicity and need. Villages may be particularly isolated, transport and infrastructure may be damaged, roads impassable and supply chains difficult to establish. Immediate help can be patchy and dependent on the area, access and established channels of governance. If any of these factors are fragile, then people may not receive relief. One man described how there was major news coverage but very slow relief. Others describe the frustration of not receiving basic support, often because of the isolation of the village and the destruction of roads and other infrastructure, making access difficult or impossible for relief agencies:

> Different TV channels came to cover us. We told them [about the cyclone]. They reported this. Still we are not receiving aid. The District commissioner, Ministers, MPs, they all came. TV channels broadcast our interviews. Don't the citizens watch it? The whole of Bangladesh watches. We were news in the papers for 15 days in a row. Yet after all this, we do not get any help, then it must be our [bad] luck!
>
> (Older man, Gaibandha)

> We did not get a VGF Card during the last flood they took our name when my house was destroyed. They said I shall get 3000 taka, rice and pulse. But they did not give us anything. We got only 10 kg rice.
>
> (Young woman, Gaibandha)

> We did not receive external aid in time as we had river all around us. People coming in with support gave away the support in other areas – they did not come into the remote areas as it is difficult.
>
> (Older man, Satkhira)

NGOs and disaster responses

While governments are engaged in disaster preparation and early warning strategies, NGOs are more likely to be part of the recovery phase of a disaster than the preparation:

> In disaster language, what we say is, 'You have rescue, recovery and then you have rehabilitation'. So, we move into the rehabilitation phase, unless of course the incident happens during our presence there … So, that means when we enter or penetrate an area, we always go with a three-pronged attack. So, that would be microfinance, it would be health and it would be education … Gender is cross-cutting, so it's always there somewhere.
>
> (NGO Key Informant, Dhaka)

Nonetheless it is clear that the NGOs have had a critical role in reducing the impacts of these major events on villages. Disaster preparedness is much improved and the immediacy of the assistance has protected people and saved lives:

> Yes, [Cyclone *Mahasen*] was a year before Bangladesh's independence. That cyclone hit on Thursday, November 12, 1970. There was massive flooding. More people lost their lives in that cyclone than this flooding (*Sidr*). Crops were washed away.
>
> (Focus group, teachers, Barguna)

> The situation is better now than that of before, because we are getting more help now. During the '70s flood, so far as I can remember, we got only some flour and that was the first time I was introduced to flour!
>
> (Older man, Barguna)

Long-term reconstruction

It is the task of government authorities to re-establish infrastructure – dams, roads and houses – following disasters. These big infrastructure projects may take time, but are essential to restoring the community, re-establishing livelihoods and building resilience.

> Government firstly reconstructed the broken dams properly, secondly the roads and houses are strongly constructed and alleviated, people are preparing for planting and [thus] the people can stop cutting the trees of the jungle [for reconstruction and as a livelihood strategy].
>
> (Young woman, Gaibandha)

While these tasks are undertaken by government instrumentalities, it is the work of the NGOs that ensures that people survive. NGOs are essential to long-term reconstruction efforts – without them the scale of the destruction would be unmanageable. There are many national and international NGOs working across Bangladesh providing ongoing assistance to communities affected by climate events. This is recognized by the people of Bangladesh – many are aware that the NGOs have saved their lives:

> The NGOs helped us a lot beside the government. They gave us a lot of tin to rebuild the houses. Also they gave us food and blankets. There is a NGO named X that gave the widow women 2,000 taka each as a grant. And gave to the others 1,000 taka each as a loan without interest for anyone who wanted to start a business or cultivate land.
>
> (Young man, Barguna)

One NGO employee noted that they are experimenting with the design of houses to try and make them cyclone-proof in order that people could remain at home:

> We've just started a pilot with limited funds trying to understand whether we can build disaster- or climate-resilient housing as an alternative model to the traditional houses that are being built – or the traditional [reconstruction] methods that the government practice … We are going to build an entire village with a school cum community centre because, if these houses are cyclone proofed or let's say, storm proofed or climate proofed, then you won't really require a cyclone shelter anymore.
>
> (NGO Key informant, Dhaka)

This idea was expanded on at a workshop in Dhaka where NGO and government representatives met to discuss our research. Discussion proceeded on how more effective designs for shelters could make them more accessible for women:

> A national committee has revised the design, and now we have two new designs accommodating the positioning of toilets for males and females in two different corners, and perhaps in a three-story cyclone shelters, a separate room for lactating mothers, for their privacy, even during cyclones.

There were many stories of successful government and NGO aid projects designed to rebuild infrastructure and restore access and support to isolated communities. One woman describes government support provided in the reconstruction phase to restore infrastructure and thereby assist people to get on with their lives. As one NGO representative explained, they have been seeking advice from local people on the types of support they require. Women in one village have requested a bunker be built to store their goods and valuables during a disaster:

> They were asking for some very, very basic necessities; they said, 'Build us a bunker.' And many people said, 'Why do you want a bunker?' They said,

> 'No, no this is where we would store our food, water and valuable documents. Once the flood and all passes, the water recedes, we can go back.'
>
> (NGO Key informant, Dhaka)

People in the villages appreciate these NGO efforts as they have dramatically changed lives:

> In the past people were ignorant, there was no NGO to motivate them. But now, every body motivates us. People are educated now. Children of the ignorant people are also going to school now.
>
> (Interview, female, Gaibandha)

NGOs report that there is a problem associated with the continuity of successful programmes. They identify a gap that could be filled by government in taking over and managing programmes such as shelter management:

> [The NGO identifies village-level change agents – like a teacher to run community meetings.] But once the project is stopped, the impact is finished. But these people will stay there forever and we don't pay even one dollar. The people were trained in shelter management, people are trained in income generation so that the shelters are managed, maintained properly. They were trained in first aid, after the cyclone. They were trained in accounting, early warning, you name it.
>
> (NGO Key informant, Dhaka)

Maintaining and expanding successful programmes remains a problem identified by many stakeholders including village-level participants:

> It is not possible for the NGOs to do all the work. For example an NGO has installed two water filters here while you might have a need of 20. An NGO might give tube-wells to 10 families out of 100. So the development initiatives have to be taken by the government. Moreover, the NGOs bring money from foreign countries – but in the case of financial recession donations decrease as well. There are other problems too, as the NGOs have to work in coordination with the local stakeholders. For example, my NGO wanted to dig a canal in [village] area, but the local leaders were against it. So often the NGOs have to step back to maintain coordination with the local community stakeholders. They leave it to the government authorities.
>
> (Older man, Satkhira)

Lack of support during slow-onset events

Despite these major undertakings, there is a distinct difference between the active engagement of government and NGOs in reconstruction after a catastrophe as opposed to action following a slow-onset event. People affected by slow-onset events note that help is not forthcoming in the same way. Slow-onset events are just as devastating but are not supported by government and NGOs:

> Listen, if the flood comes and our house is drowned, our relatives and neighbours come and help us. But during drought we do not get any help from anybody. Do you get me! We do not get anything at that time. Suppose, a strong wind or storm destroyed my home or some people get injured, somebody will come and help us. But for drought, nobody is there.
>
> (Older woman, Gaibandha)

In Gaibandha, 72 per cent of respondents noted they did not receive any strategies or assistance from an NGO. The most successful strategy reported by respondents was NGO provision of microcredit and domestic animals. Other strategies included discouraging the exchange of dowry and early marriage, encouraging education, distributing seeds and rice, and giving information about weather and disasters.

In Barguna, 75 per cent noted that they did not receive any strategies or assistance from an NGO. The most successful strategy reported was the giving of microcredit or loans. Other strategies included rebuilding houses, distributing seeds, rice, food and utensils, providing information on animal rearing, cultivation and other income-generating work.

In Satkhira, 76 per cent noted that they did not receive any strategies or assistance from an NGO. The most successful strategies reported were receiving microcredit or loans, being given soil-filling work and the distribution of water filters. Other strategies included rebuilding houses, the distribution of seeds, rice and food, the provision of information on crab-farming and cultivation, and the building of toilets and cyclone shelters.

Survey participants were asked to comment on the adequacy of government support for those affected by slow-onset climate challenges. Most (80 per cent) suggested that there is a lack of government support or assistance. The vast majority felt no help was forthcoming. This was highest in Gaibandha (84 per cent) where respondents noted that they had had no help. In Barguna it was 71 per cent; and in Satkhira, 81 per cent. When asked what assistance they would like there were a range of responses including information about family planning, gender equality, agriculture, how to obtain food, how to access money to rebuild a house, allowances, child vaccinations and loans.

NGO principles for action

NGOs are crucial to post-disaster work and to ongoing development. Yet the significance of their work lies not just in the direct aid and training that they provide but also in the formulation of principles for action. This is a critical factor as the principles that guide their work are absorbed by villagers and can shape future actions and activism. Human rights and gender equality are underlying principles for many NGOs.

Larger NGOs have now organized themselves into an informal lobby group to raise issues with government and to lobby for major changes that will assist them in their work. One area where there has been widespread cooperation is in the area of women's rights:

> come together constantly around the campaign for women's rights. The recent campaign was around domestic violence legislation. It has been adopted and it is now an Act.
>
> (NGO Key informant, Dhaka)

Several organizations note that they are moving to a rights-based approach rather than merely giving aid. In this way they hope to leverage their limited resources to assist individuals and communities to grow. This view was echoed by several NGO key informants and foreshadows that significant change is possible through post-disaster and community development approaches:

> The first 10 or 15 years [our aid] approach was a very direct engagement, service-oriented approach and if you go to some of those parts of Bangladesh you will find we've built schools, we've sunk tube bores, we've lived there, worked there and that sort of approach. But later on, it became evident to [NGOs] across the globe that we are looking more for rights and entitlements and empowerment. So how do we work on those areas? [We need to build] people's capacity, [to] harness it. And then our approach to our whole work became a very rights-based approach. We're not a rights-based organisation; we're not doing human rights campaigning, but we have a human rights-based approach. And we are very, very clearly, taking sides with the poor.
>
> (NGO Key informant, Dhaka)

> Whatever we do is rights based. It's not, giving something or helping something or even giving some materials, nothing.
>
> (NGO Key informant, Dhaka)

> We don't see the problem of disaster in isolation from the problem faced by the poorest of the poor. Our focus is not narrowed down to either climate change or disaster risk reductions. Our focus is empowering community so that they themselves can identify the problem and they can identify the risks, they can identify the root causes of the risks and they can treat those issues.
>
> (NGO Key informant, Dhaka)

The move to a rights-based approach, and the focus on women's rights, is noted at village level and has been quietly applauded by local women. There is little doubt that once seemingly immutable gender relations are open to challenge through well articulated equality principles.

> Because of the training organized by various NGOs our family members especially the men started to help us in managing some household chores.
>
> (Older woman, Barguna)

Layered government support

The Bangladeshi national government is providing an increasingly effective, targeted social safety net that has lifted millions out of poverty:

> Over the last two financial years, our social safety net programme increased, compared to eight years ago, increased from say, 6 to 7 per cent. Now it is 33 or plus percentage. … But a significant proportion of 4.2 billion dollars, a significant proportion of that is going for the poorest of the poor. Now, whether or not they are getting the services, that's a different story.
>
> (NGO Key informant, Dhaka)

Assistance to individual households is also very much dependent on local-level layers of governance and organizing. Strong governance structures at this level are crucial to how effectively aid and resources are distributed:

> We received rice and also the chairman and members helped us a lot. The NGOs helped us after just one day of the disaster while the support from the government arrived a little later on.
>
> (Focus group, Satkhira)

Stable, fair and representative decentralization of disaster responses to the local levels of organizing can draw on local knowledge and assist to develop ongoing mitigation efforts:

> There is an administration committee, a relief committee and so on. Basically the members of these committees are the same people. If the committees could have been decentralized, that would have been useful. They need to keep working to raise public awareness so that they take up different preparedness measures such as homestea- raising before a disaster, preserving of dry food, planting trees etc.
>
> (Focus group, Satkhira)

Well-organized local level governance, even if inactive during stable periods, can ensure comprehensive attention to reconstruction during a crisis:

> I had attended a conference on disaster management in Dhaka. Once I returned from there, I gave the union parishad Chairman a note on disaster management. After that the chairman formed a disaster management committee. I was a member as a farmer representative. Before *Aila*

> basically the committee was inactive. It was formed so that we could get an idea of what to do in case of an emergency. All the members of the union parishad took the initiative jointly. It was basically formed to ensure that the community members have defined roles in times of a disaster. Like in case of disseminating the early warning through a hand mike, taking people to shelters, ensuring proper distribution of relief aid, etc. Then they have the roles throughout the rehabilitation phase as well. To ensure community participation, people from all levels of the community are taken into this committee.
>
> (Older man, Satkhira)

While local-level organizing is crucial, there are still gaps between NGO reconstruction actions and local-level organizing and control. It is in the murky waters of layers of governance that efforts to ensure positive adaptation can fail:

> All the 3,000 cyclone shelters that have already been built, there is hardly any sense of ownership of the common people. They had no say in where the shelter should be built, who is going to manage it, who will have the key.
>
> (NGO Key informant, Dhaka)

One NGO representative also noted this failure, arguing that their attempts to organize from the local level to national actions have not been successful because of issues of governance at higher levels:

> Based on our experience in the field, we organized a national-level advocacy program with the Ministers and the Directors General of key organizations, bringing in people from the community, so these people can give them knowledge 'straight from the horse's mouth' so to speak. It was very well appreciated by a lot of people. When I came back, the Director General of a very important division of government, called me, saying that, 'Your colleagues are creating a problem in our society.' I had to explain to him that this is the only way, if you want to have any sustainable change in the whole system, otherwise, you know, the trickle-down theory, doesn't work from here. … we are trying to bring this voice from the community not only to national level but also international level.
>
> (NGO Key informant, Dhaka)

Corruption

Corruption in the distribution of resources following a disaster can create significant hardship for women. Under customary law it is usual for a male household member to seek relief. Women are therefore dependent on others to receive help and this may lead to corruption of the processes of distribution and significant gender differences with regard to who benefits. Nasreen's

(2008) study found that only 52 per cent of the women in Gaibanda and 59 per cent in Faridpur received government relief, due to corrupt officials, unjust distribution preference and a lack of a male relative to receive the relief. Furthermore, men received more relief than women and nepotism exists in relief distribution. Harassment of women also occurs in relief queues (WEDO 2008; APIT 2009):

> Corruption has impacted women in different ways because they find that it's difficult for them to go and write their name down to get relief when there's relief distribution. And they have to request someone from the group/family [to get their aid supplies]. ... we make it a point that at least 50 to 60 per cent of what we are giving goes to women. But the distribution list is not necessarily prepared by the NGO. The local authority, the politicians on the ground, they [decide who gets relief], so there's a tension.
>
> (NGO Key informant, Dhaka)

Summary

I have articulated four stages of disaster recovery – the preparation and immediate experience of the event, the immediate aftermath and the long-term process of reconstruction. What is evident from our findings is that people are often unprepared for the scale of the damage done by the event, that they have limited warning and that catastrophic events often wreak havoc with people's lives. This chapter illustrates the importance of government and NGO support and highlights the need for clear governance pathways between layers of government and between NGOs and government.

The Bangladeshi government has responded to catastrophic events in recent times by introducing early warning systems and establishing shelters across the regions. These shelters are thinly spread and do not have the capacity to protect all residents. Further, there are gender differences in relation to those who seek shelter. Women are less likely to receive early warnings, are more dependent on men to escort them and are more reluctant to attend shelters because of the threat to their honour and comfort. While governments are more engaged in providing early warnings and preparing people for disasters, NGOs are more likely to act in the immediate post-disaster phase. What is clear from this research is that there is extensive support during and after a catastrophic weather event but very little similar strategies for those experiencing slow-onset events. In this chapter we present our findings on the impacts of these climate challenges on the people of rural Bangladesh.

7 Impacts of climate change

In this chapter I examine the impacts of catastrophic and slow-onset events on the people in the three regions of Bangladesh that form the basis of this study. Moving away from the *perceptions* of climate change, I examine the evident *impacts* on local people and communities. Given the list of potential climatic changes detailed in Chapter 6, I outline the way in which these have affected livelihoods, customary practices, education, health, migration, agriculture, food security, relief, water and environment. I note the way these impacts challenge existing customary practices and open up the potential for more radical actions that incorporate gender equality measures.

Cyclone *Sidr* occurred in 2007. This interview took place in 2011, and yet the experience is still raw, recollection immediate and long-term health consequences evident:

> The water [from Cyclone *Sidr*] came at 9pm. We swam from 9pm until midnight then went to the tree. The whole night we spent in the tree naked. I only had a blouse and petticoat, no sari. I took off the sari while swimming. – I had to save my life. From then on [I have been sick with a] cold. I didn't know how long it would take to get well. If I bathe one day then next day I cannot [I am too weak].
>
> (Young woman, Barguna)

How do these events impact people's lives, their ongoing livelihoods and well-being? First we examine the very personal losses recorded by respondents to our survey. We then address the impacts on livelihoods and agriculture, reconstruction, health, education, traditional customs, and gender relations.

Impacts of climate variability

Livelihoods

Survey respondents were asked to comment on the obvious impacts of climate/weather events on their families, land, houses, livelihoods and family arrangements during the last ten years. Table 7.1 provides a summary of

Table 7.1 Numbers of respondents who noted particular losses

Type of loss	*Women*	*Men*	*Total (Gaib/Barg/Sat)*
Loss of family member	6%	5%	5.5% (10/11/5)
Injury to family member	13%	21%	17% (45/35/26)
Loss/damage to house	78.5%	65%	72% (145/149/139)
Loss/damage to crops	73%	65%	67% (154/155/105)
Loss/damage to garden/nursery/ vegetables	68%	47%	39% (86/95/56)
Loss/damage to livestock	56%	45%	51% (115/124/74)
Damage to fodder	19%	22%	20.5% (53/47/27)
Loss/damage to pond	17%	17%	17% (28/40/39)

responses. A significant proportion (over 20 per cent) noted the death or injury to a family member, 72 per cent loss or damage to their house and 67 per cent loss or damage to crops. Injury to a family member, the loss or damage to crops and fodder, and out-migration (predominantly of men) are highest in Gaibandha, while loss of life was highest in Barguna, where Cyclone *Sidr* was most destructive. Not all households have access to ponds for their water needs. Many households are reliant on water from rivers, requiring (usually) women to walk long distances several times a day. It is therefore noteworthy that nearly 20 per cent of respondents stated that ponds have been damaged – this would represent a significant proportion of available ponds.

The people who reported these significant losses expressed deep devastation at the scale of their personal experiences. It is difficult for people to discuss these losses that impact so heavily on their lives. NGO representatives also express despair at the scale of the damage:

> We work with communities who live there. Their houses are gone, their food security is gone, the water source is disturbed, children cannot go back to school, health is affected, and some of the areas where we [work], there's none of these services anyway, especially in the river islands called the Chars.
>
> (NGO Key informant, Dhaka)

There has been a laudable reduction in the loss of life from disasters – from the time of the 1991 Cyclone *Bhola*, when 130,000 people died, most of them women. During Cyclone *Mahasen* in 2013 only 14 people perished. This has resulted from the significant efforts made by governments and NGOs to prepare for and address the immediate post-disaster recovery phase. Nonetheless over time it has become clear that other losses are critical to the ongoing health and well-being of the people in the direct path of major weather events. These critical catastrophic events are having a direct impact on the ability of people to re-establish their lives. Immediate and ongoing factors following a disaster include displacement, infrastructure losses – including roads and transport – livelihood disruption and long-term impacts on food and water security.

> The government didn't want to recognize it as an emergency because there were only a hundred or something people who died. But the fact that it displaced so many, they lost their social income, they didn't have a place to stay – that sank in like a year later. And then this area, it's very difficult to reconstruct the roads in the traditional ways. The river current and erosion is like an everyday occurrence – you finish a bit of the construction, you come back the next morning and it's disappeared. So, it's been quite tough and that remains a problem.
>
> (NGO Key informant, Dhaka)

The construction of houses has also been delayed:

> At Satkhira, … you will see what *Aila* has done. It's a place close to the mangroves area – the Sundarbans. That's the largest mangrove forest in the world …. There's still people… living under the open sky. Some are building their houses on their own plots; some are building theirs beside the road. Those who have a stable piece of land – they are rebuilding their houses on it.
>
> (Young man, Gaibandha)

The government also appears powerless to deal with the scale of the emergency which covers such a significant proportion of the country and has affected so many millions of people:

> No, we're not in a position to provide everyone with shelter. Initially when the first shelter was constructed …. There was one room where everybody used to stay. But now the people do not want to leave their houses, leave their cattle and horses in danger because if they go, then they will be in very serious trouble. Now we have decided that all of the flood shelters will be redesigned. The ground floor will be vacant because of storm surge. First floor, we have provided a ramp up to the third floor so that disabled or old people can move easily. And the first floor there will be livestock, cattle. The next floor there it will be children and women and disabled persons, and in the highest floor, it will be the people who are staying there, and we have provided a map for them.
>
> (Government representative, Dhaka)

In Gaibandha, riverbank erosion is causing huge chunks of farm land to fall into the river, a phenomenon we witnessed during our visit. It was explained to us that the land is diminishing at a rate of approximately half a kilometre per year in some places. This is not only reducing the size of productive farm land, it has also resulted in land becoming worthless. Although not a precise conversion, the following farmer has lost all but under an acre of his previous farm of approximately 22 acres:

> We can't sell our lands. The river erosion is taking everything away. My father was the only child of his parents. And we are two brothers. Our father had 22 or 24 bighas[1] of land. But now we have only lands measuring about 7 or 8 shataks.[2] Everything else is taken by the erosion. Think about the present and the past. What we had then and now what we have.
>
> (Interview, male, Gaibandha)

Impacts on livelihoods by district

Respondents were asked to comment on additional impacts on their occupational status resulting from climate events. These varied somewhat by district. Nonetheless, participants note that they must make a number of changes to their livelihoods in order to earn enough to survive. While the top responses to our questions concerning the impacts on livelihoods were 'no change' or 'do not know', constant responses across the three regions were i) a need to change occupations, ii) a need to do more work, iii) that there is no work available and iv) that health issues are impacting capacity to re-establish livelihoods.

Table 7.2 summarizes these responses by region.

Other impacts on occupational status noted by participants included that women now work outside, that women need to undertake paid work to cover household expenses, that there is a growing need for children in some families

Table 7.2 Summary of occupational impacts by region

Type of impact	*Gaibandha*	*Baguna*	*Satkhira*	*Total*
Need to change occupation	13%	11%	15%	13%
More work	9%	8%	3%	7%
Lower income	8%	6%	7%	7%
Loss of business	2%	1%	3%	2%
Work more difficult	8%	10%	9%	9%
No fish in the river	8%	2%	3%	2%
Lack of work	5%	6%	2.5%	4.5%
Damage to infrastructure affecting access to work	4%	1%	3%	3%
Difficult to do work	3%	3%	5%	4%
Physical problems	5%	5%	7%	5%
Lack of food or water	1%	5%	2%	3%
Do not know	15%	12%	9%	12%
No change	20%	23%	25%	23%

to leave school to work, that there is great difficulty associated with working during and after a disaster, that there is a need to find a variety of jobs at the same time, that people feel sick most of the time and that it takes time to repair the home after it is damaged.

Impacts on agriculture/cultivation

There have been changes in crop cultivation in all areas. In Gaibandha people noted that weather change and major flooding has reduced the number of trees and increased the amount of sand in the soil. Farmers have moved from rice and lentil production to peanuts, corn and rice. In Barguna farmers have move into sunflower production, while in Satkhira farmers are using the salt water intrusion into their once-productive soils to create shrimp farms.

In all areas people note that the temperature has increased. In every area it is obvious that the suffering and anguish of women has increased as they cope with food and water security issues for their families. Women consume less food in all areas and are thus more vulnerable to diseases and illness. Because they are dependent on men, they have no ready access to medical facilities without permission. A further very evident adaptive strategy in each area is that women are responsible for household activities as well as income-generating work.

The impact of climate change on agriculture in Bangladesh is of immense significance not only because of the threat it poses to food security, but also because the increased risks of malnourishment, seasonal hunger and stunted growth are very real. Two thirds of Bangladeshis work in some form of agricultural production, and the threats to livelihoods and ongoing poverty reduction strategies posed by climate change are critical. Some of the adaptations adopted by Bangladeshis include changing crops, modifying cropping strategies, introducing new technologies and sourcing income from other activities.

Reduced yields

Participants in the interviews and focus groups note that yields have gone down due to reduced rain and higher salinity and that this is having a significant impact on food security:

> I used to cultivate my land thrice in a year but now I cultivate only once because now we don't get rain.
>
> (Male focus group, Barguna)

> The lands used to produce two harvests a year before [Cyclone] *Aila*. But since *Aila*, it has become good for only a single harvest.
>
> (Male focus group, Satkhira)

> Crop [yield] has been reduced. Before we didn't have that much salinity, which now has increased [greatly]. Farmers face difficulties in cultivation

now – they can't rear animals (cows, buffalos), because of the scarcity of good animal-food and this lack of quality animals [to pull planting equipment] causes difficulties to cultivation.

(Older man, Barguna)

The crops we grow in this area, such as lentil, pepper, various types of vegetables – their production has fallen sharply. Salinity in the cropland has increased … Even if they survive the yield is lower. Yes, this is a cause of the sufferings of the farmers of this village. We have been noticing these sufferings in [relation to the] cultivation of our crops. This is the time for cultivation of vegetables, but if you go to see the situation for yourself, you will find that there are almost no vegetables in the village.

(Women's focus group, Barguna)

We cultivate potato in our area … Because of salinity the production has decreased.

(Male focus group, Barguna)

Fruit tree damage

Village people traditionally rely on fruit-bearing trees to supplement their food supplies. However, increasing salinity and flooding are destroying these trees:

Mendi or chambol, mehegoni or coconut – [we have less of] these types of trees. These do not give a good amount of fruits. We used to get 10/20 bags of mangoes (50/60 kari) when my parents or grandparents [were farming] but these days mango yield has been decreased significantly. The same is the case for jackfruit. When it needs rain, it is absent.

(Older male, Barguna)

There are many trees around my house … the yield of the betel nut trees has fallen sharply. These betel nut trees look healthy, but nothing is growing in these trees. My jackfruit trees are about to die. Some have already died after *Sidr*. Some other trees – such as mahogany, chambol, are not growing properly.

(Male focus group, Barguna)

[We used to] plant trees – they would produce fruits then poor people can live by selling the [fruit]. After *Sidr* these have come down. Coconut trees sometimes grow, sometimes not. Betel nut, before it would grow 3 times more, now it is one-third. Papaya does not grow … it is the saline water. Many things have decreased. Now you see there is no peace without suffering! The paddy is not okay. Because of salinity the crops die, they get insect infected, the rain is coming quite often, the roots rot, the flowers

> fall, the tree becomes red, the leaves become white … Only Allah knows, people don't know anything.
>
> Children suffer a lot [because of] food [shortages]. We can't even collect vegetables for the children. If we have any income then we can buy something from the bazaar. Otherwise [there is] totally nothing.
>
> (Young woman, Barguna)

Changed production techniques

Farmers have introduced new production techniques including insecticides to control insects and a much heavier reliance on fertilizers – this has a critical impact on ground water – in many areas the source of village drinking water:

> Before *Sidr*, we used to apply half a kilogram of fertilizer, now for the same land we have to use one and a half kilogram to two kilograms of fertilizer but even that doesn't work. We can't plough our land properly. Due to the lack of a canal we [can't] arrange irrigation. We try to irrigate our land by taking water from our roadside where water is stranded.
>
> (Male focus group, Barguna)

> As the fertility of our land has decreased, so we are using urea and other chemical fertilizers. Due to continued use of chemical fertilizers our land has lost its fertility. For example, in a piece of one decimal of land 200 grams of fertilizers were required earlier, but now we need to use 500 grams for that piece of land. Earlier, grass and other non-crop plants used to grow abundantly in the field, but no non-crop plant grows these days.
>
> (Male focus group, Barguna)

> Winter crops grow very well in our area. But to increase the fertility of the land, we need to use some fertilizers. We produce those fertilizers. In our village, only cow dung was used as fertilizer. Farmers did not use any other fertilizers. As the fertility of our land has decreased, so we are using urea and other chemical fertilizers. Due to continued use of chemical fertilizers our land has lost its fertility.
>
> Moreover, we also use an insecticide. If we use insecticide during the planting phase, then the plants will become bitter-tasted. This wards off insects from the plants. As a result, whatever we grow – pumpkin, gourd etc. – these remain in these crops … The insecticide people use in cultivating vegetables remains in the vegetables and thus we consume those. Insects cannot do any harm to these vegetables due to the presence of insecticides, but when we people eat those vegetables, we suffer from various types of diseases.
>
> (Male focus group, Barguna)

> Yes, it needs fertilizer, [and in the] dry season it needs irrigation and pesticides … this is now a place of one crop a year.
>
> (Older man, Satkhira)

Changes to the types of crops grown

Because of climate challenges, and in an attempt to second-guess the weather, farmers are experimenting with growing different types of crops at different times of year:

> This year the flood came very early – that's why cultivated paddy has been washed away. People could not harvest the paddy. Crop fields are cultivated 2 months earlier. We could not plant rice in time. Therefore, we had to plant corn 2 months earlier in the paddy field.
>
> (Male focus group, Gaibandha)

> It has been three years now since it started to change, since [Cyclone] *Aila* to be specific. It was all good before the cyclone. We used to cultivate paddy, jute and other crops. But since *Aila*, the rain and flood have become irregular.
>
> Currently we are cultivating [several different rice varieties]. Previously we used to cultivate [different ones]. Importantly, the species cultivated now are expensive but give a better harvest … though more labour and cost are needed … but we selected it for increasing the volume of production.
>
> (Male focus groups, Satkhira)

> But there was much virus. Pesticides aren't working; again more pesticides are harmful to crops. In some places salt is harmful. After *Sidr*, until now. Before *Sidr* Pamuri (insect) was seen. Now it is not seen. But now different types of unknown insects are seen.
>
> (Older man, Barguna)

Impact on food security

The impact on food security is significant. Respondents note their reduced ability to grow food, reduced availability and quality of fresh food, an increased identified pressure to earn more income for food, that there is less variety of food and, if they must purchase food, that they can only afford to buy rice. Table 7.3 provides more detailed information on food security issues.

Food insecurity is causing a significant crisis in families and many participants refer to this as their 'suffering':

> We suffer. If everything is flooded in the water then [we have difficulty] cooking, collecting, we have to go to and fro for cooking, collecting water

Table 7.3 Food security factors

Type of issue	*Women*	*Men*	*Total*	*Comment*
Pressure to earn more income for food	82%	75%	78%	More women than men raised this issue. Highest in Gaibandha
Reduced ability to grow own food	75%	65%	70%	Gaibandha 74% Barguna 68% Satkhira 68%
Reduced availability of fresh food	32.5%	41%	37%	
Less variety	36.5%	50%	44%	
Reduced quality of fresh food			25%	

> and feeding the children. Isn't it a suffering? … Mothers remain occupied with these types of suffering. Children cry because of hunger … Women and children suffer the most.
>
> (Interview, female, Barguna)

> It has affected us all very badly. In our area, or in our markets we do not get these [foods] … In this village, vegetables are almost unavailable. This is the situation.
>
> (Focus group, male, Barguna)

> Earlier we had milk, butter and ghee in plenty to eat. But now our children have only heard of them. All because of saline water. The cows and goats are not getting straw to eat, because paddy cultivation is becoming rare day by day.
>
> (Female focus group, Satkhira)

> According to one of my neighbours [her son] is eating only once a day, while before he used to eat 3 times in a day. [That] boy comes to school after having watered rice but he is not getting proper vegetables to eat with that rice. He needs water but it is salty. Most of the people are suffering for lack of vegetables. They aren't getting a proper diet.
>
> (Older woman, Barguna)

Even when families do produce food they may be forced to sell this in the markets to generate income. This has impacts on food insecurity within families – often it is the women who suffer most:

> Even if they grow two kilograms, a lady will ask her son to go to the market, sell it and bring home rice or something like that. That's how it is. They rear poultry. Often they don't get the benefit of it because, it is

> for the urban market: … even within households, where food is scarce, women will eat less rather than sharing out [what they have] … Women eat the last portion, no matter how small it is and perhaps the worst quality food.
>
> (NGO Key informant, Dhaka)

Food insecurity, and the impact of rising food prices, is having a significant impact on nutrition in rural areas and is resulting in an unfair distribution of food within the country:

> We produce only 11 per cent of our total edible oil demand in country – the rest is imported. That means that those of us living in urban areas, who have a good healthy income, we don't have any problems. But with changing and fluctuating prices, the poorest of the poor, I think more than 50 million, they don't have enough edible oil, vegetable protein, lentils and other things – they can't have them because it's pricy and it's imported.
>
> (NGO key informant, Dhaka)

Other comments on food security are more blunt:

> The government refuses to call it famine, so [it is] scarcity of food – it's a drought-prone area and [there's a] lack of sufficient rain and ability to cultivate. This has been going on for 40 or 50 years. But the intensity is increasing now. They're losing more land to river erosion. And there, the girls, at a much younger age, there is child marriage, and there's also trafficking.
>
> (NGO Key informant, Dhaka)

Impact on water and fuel security

Survey participants noted that water security is compromised by the reduced availability of fresh water, the need to find new sources of drinking water, the increased time it takes to collect water, and increasing salinity affecting water supplies in the dry season, particularly in coastal areas. Women note that they often walk through contaminated water for hours to reach fresh water, leading to significant health impacts including skin diseases and gynaecological problems. In Satkhira the impacts on water security are increasingly being felt by men as the contamination of water sources, and the long distances to new sources of water, has necessitated men helping with this task. The longer distances women must walk to collect water is also challenging traditional restrictions on women's freedom of movement. Table 7.4 provides a summary of responses relating to water security.

The changes relating to fresh water access have particularly insidious health effects. For example, apart from the greater distances women now must go to

Table 7.4 Impact on water security

Water security issue	*Women*	*Men*	*Total*	*Comment*
Reduced availability of drinking water	63%	66%	64.5%	Satkhira (80%)
Increase in water availability	6%	9%	7.5%	
Salinity in dry season	54%	56%	55%	Satkhira (96%) Barguna (69%) Gaibandha (4%)
Increased time to collect water	47%	46%	46%	
Need to find new source of drinking water	48%	44.5%	46%	Satkhira higher
Change in who collects water	14%	16%	15%	All changes involve men becoming more involved

collect water, often walking through contaminated water along the way, they are also training their children to drink less – a process that has ongoing and serious health consequences. Nonetheless there is some evidence of interventions from NGOs to assist with water-purifying and filtering in a number of villages:

> Women walk several kilometres to get a pitcher full of water. And they train their offspring, that's even more painful – they ration water. Because otherwise, on a single day, they have to go there again, walking [through contaminated] water, walking there and coming back. Because there is no system for them to be able to collect it in their dwelling …they train their children to drink less. So, that risks their organs – because there is no alternative, right. Otherwise they are forced to drink the same water that they defecate in.
>
> (NGO Key Informant, Dhaka)

> Sometimes we don't even get water to drink. Sometimes the tube well goes under the water. Once we drank the water during a flood, and faced serious misery – fever and coughs.
>
> (Older woman, Gaibandha)

> We used to drink pond water here. The pond water was pure. Everybody here used to drink pond water. When I became involved with the NGOs, I was the president of water aid. I advised village people not to go to the toilet without shoes and [to] wash their hands … [With the help of an NGO] I called a meeting to discuss obtaining a water purifier and one

> [villager] said, he will give the land to set the filter if the people get benefit. The [NGO] sent their engineer and I placed an application to them. They set a filter here and people get drinking water. Earlier there were diseases like dysentery from drinking pond water. Now it has been decreased.
>
> (Interview, female, Satkhira)

There have been significant impacts on lives and livelihoods resulting from reduced availability of fuel for cooking, the need to find new sources of fuel and increased time taken to collect fuel. Ninety per cent of respondents noted that there was less fuel available (and more women than men); 73 per cent noted the need to find new sources of fuel; and 36 per cent noted the increased time needed to collect fuel (this was more pronounced in Satkhira).

Impacts on education

There are a number of impacts of climate events on educational access, including the increased time to get to school, pressure to take children out of school, poverty and deprivation resulting from trying to keep children in school, and early marriage for girls (see Table 7.5).

Of the 58 people who took their daughter out of school for an early marriage, the age at her marriage was between 12 and 17 (see Table 7.6). Child

Table 7.5 Impacts on educational access

Type of impact	*Women*	*Men*	*Total*	*Comment*
Increased hardship in getting children to school	22%	24%	23%	Poor roads and lack of social security
Pressure to take children out of school	8%	10%	9%	
Financial hardship	92%	90%	91%	
Took daughter out of school to work	14%	14%	14%	most of these in Barguna
Took son out of school to work	9%	8%	9%	
Gone without something/ sold something to keep daughter in school	7%	11%	9%	
Gone without something/ sold something to keep son in school	10%	11%	11%	
Took daughter out of school for marriage	9%	8%	9%	

Table 7.6 Age at which girls were married as a result of climate events

Age – yrs	*Number*
12	9
13	22
14	7
15	9
16	7
17	4

and forced marriage is a direct consequence of climate events and consequent poverty and hardship.

The impacts of climate events and landscape degradation on families, their homes and livelihoods make accessing education for children a very difficult proposition:

> Because of erosion, people have lost their house, land, property. They have to stay in other people's houses. They are unable to eat/have three meals a day. So it is quite impossible for them to support their children to continue study. This is the main problem.
>
> (Young male focus group, Gaibandha)

> Many of us have dropped out due to the financial crisis in our families.
>
> (Young female focus group, Satkhira)

Impacts on health

Various physical and mental health issues are associated with natural disasters. For example pregnant women in shelters are at greater risk of abortion and miscarriage, and childbirths are handled with minimal privacy and lack of care (Ahmed *et al.* 2007; Save the Children 2006). Pregnant women also have a higher risk of malaria (Mitchell *et al.* 2007). Access to sanitary latrines is difficult during a climate crisis (Nasreen 2008; Kumar and Narayan 2007; World Bank 2006), and very poor sanitary conditions in shelters and homes can result in urinary tract infections, diarrhoea, dysentery, skin diseases and hygiene issues with menstruation (Ahmed *et al.* 2008; WEDO 2008; Rashid and Michaud 2000). Women in saline-affected areas report experiencing gynaecological problems and menstrual pain due to saline water (Ahmed *et al.* 2008). Women, already malnourished, generally eat last or not at all to ensure food security for their family members (Ahmed *et al.* 2008) and therefore experience more health issues (Cannon 2002). Yet access to health services is severely limited (Nasreen 2008; Rashid and Michaud 2000).

Fear, anxiety and insecurity are commonly reported in shelters and at home, including fear associated with abuse and harassment, of repeat disasters, of the loss of animals and of children drowning (Kartiki 2011; Nasreen 2008; Ahmed *et al.* 2007). Further, extensive and increased post-disaster violence against women results in isolation, stigma, loss, fear and depression (APIT 2009). Women experience emotional trauma relating to social norms (such as fishing in open daylight) and because of their inability to send their children to school (Ahmed *et al.* 2007). Women also report difficulties searching for missing family members (Oxfam GB and Naripad 2011).

Health consequences are common across the three areas that formed the basis of our research. Health-related illnesses directly relating to climate and weather events include an increase in waterborne diseases, particularly diarrhoea in Gaibandha; and waterborne diseases and skin diseases in Satkhira. Common illnesses across the three regions are colds, coughs, breathing difficulties, skin problems, diarrhoea, typhoid, reproductive health issues, malnutrition, high blood pressure, dehydration, jaundice, eclampsia, anxiety and mental health issues. In all regions women note the significant health issues related to sanitation, defecation and maintaining menstrual hygiene. Over 80 per cent noted a decline in general health – slightly more men than women, and higher in the south. Table 7.7 outlines the health issues noted by participants and distribution of these by gender.

We invited respondents to nominate who is most affected. It is noteworthy that women and men are more likely to nominate women as being affected than men and that both women and men note the impact on girls and boys. Table 7.8 indicates who male and female respondents felt were most affected.

When asked the reasons for poor health, most (61 per cent) said they didn't know, 26 per cent blamed the increased temperatures, 5 per cent the lack of clean water and 4 per cent the lack of food and malnutrition. These responses were evenly spread, with people from all age groups reporting the impact of temperature fluctuation. Participants in interviews and focus groups noted the impacts on health:

> A girl works in my house like a maid servant – she has skin disease. This is called 'BIGAR' in our area. There are some students in our school who have serious skin disease. They use the saline water from the canal [to wash].
>
> (Young woman teacher, Barguna)

> During floods we suffer from cholera, typhoid, dysentery. In the dry season there is dysentery and jaundice … The diseases that are carried by water increase during flood.
>
> (Older woman, Gaibandha)

> Heart failure, fever, diarrhoea and other diseases [are present here]. We have to spend huge amounts of money for treatment … my elder sister's

Table 7.7 Health issues resulting from climate changes

Health issue	*Women*	*Men*	*Total*	*Comment*
Reduced general health	79%	82%	81%	
Increased general health	3%	4%	7%	
New health issue	52%	50%	51%	
Flu	72%	75.5%	74%	
Cold	51%	56%	54%	
Cough	50%	55%	52.5%	
Breathing difficulty	20%	32%	27%	
Malnutrition	30%	33%	32%	
Reproductive health issue	12%	12%	12%	
High blood pressure	7%	13%	10%	
Heart ailments	48%	54%	51%	
Dehydration	27%	32%	30%	
Typhoid	9%	13%	11%	61% of these in Barguna
Jaundice	23%	30%	27%	More in Barguna, less in Satkhira
Eclampsia	5%	1%	3%	Men reporting on their partners
Arthritis	7%	5%	6%	65% of these in Barguna
Skin problems	16%	12.5%	14%	Half of these in Satkhira
Anxiety/stress	5%	7%	6%	53% of these in Satkhira
Mental health issues	6%	4%	5%	52% of these in Satkhira

daughter died … our legs become rotten, all of us suffer from skin diseases. We suffer from diarrhoea and cholera.

(Older woman, Gaibandha)

Physically I was strong, but now I cannot make my hand bend for pain. There is pain in my legs, if I stand now I cannot make my legs straight for pain. The children are becoming sick with colds, cough, fever, skin diseases.

(Older woman, Gaibandha)

Table 7.8 Perceptions of whose health is most affected

Whose health most affected?	*Women*	*Men*	*Total*
Women	55%	68%	62%
Men	19%	21%	20%
Elders	46%	43%	45%
Girls	55%	50%	53%
Boys	57%	54.5%	56%
Young women	16%	21%	18%
Young men	17%	22%	20%

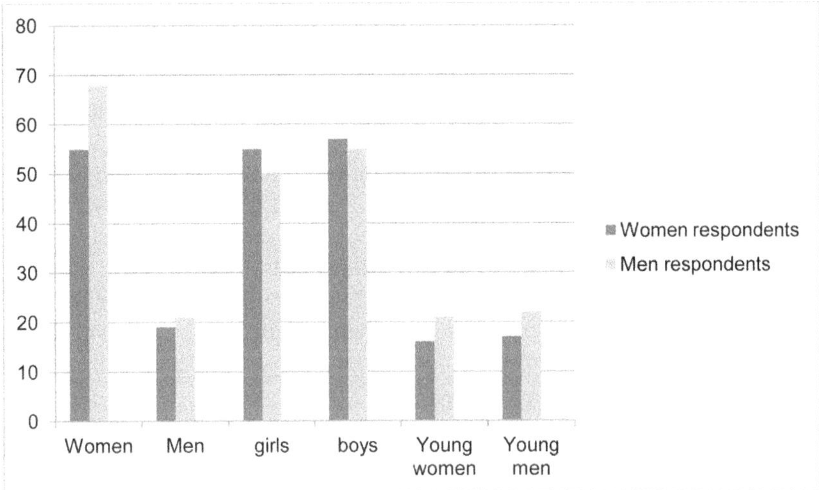

Figure 7.1 Whose health is most affected?

> Problems like itching, dandruff, allergy, skin diseases these have increased due to saline water. Cholera and dysentery are also affecting people.
>
> (Malc focus group, Satkhira)

> Different contagious diseases are spreading through mosquitoes. The saline water is congenial for mosquitoes.
>
> (Female focus group, Satkhira)

Distressingly there were a number of people who noted the impact of water and food security issues on young women and child-bearing mothers. One NGO representative noted the serious implications for women in the Satkhira region needing to walk through water every day to collect fresh water.

> We talked to different … doctors because we couldn't find any MBBS doctors over there. And they informed me that other than skin diseases – because you can see the skin diseases – at the age of 18 or 20 or 21, women are forced to remove their ovaries because there is so much infection because they cannot have any sort of sanitation. At the age of 18, 20, 21, very early. So, think about their reproductive, think about their regular normal married life. The social trauma that is going on there. It is very severe. We found out in that water-ogged area … in Bangladesh it is common that the bride and the groom never meet; it is generally the parents that arrange it. But in that discussion, when it comes, 'Okay, you are from that area?' So, no male in the neighbourhood would like to marry a girl who belongs to that waterlogged area, because she has skin diseases [and may not reproduce].
>
> (NGO Key informant, Dhaka)

> If you talk to doctors and gynaecologists you will find that among women, urinary tract infection is much higher, particularly in those regions, precisely because of the sanitation problem, it's very acute.
>
> (NGO Key informant, Dhaka)

Pregnant women's need for additional nutritious food is most often not met. It is very common for women to deny themselves food in order that their family can have more. As a result pregnant women are particularly susceptible to health issues:

> Once a girl is pregnant, we suggest she eats more. But once there are more members in the family, it becomes a problem to arrange extra food. Also pregnant women feel shy about eating more in a family full of needs.
>
> (Young man, Satkhira)

> When my wife was alive she used to work equally with me. She died few days ago. She died of tetanus. She was 7 month pregnant.
>
> (Young man, Barguna)

> [My wife died] she was pregnant, but she died of diphtheria. We spent 50–60 thousand taka for treatment. There were so many diseases after *Sidr*. After *Sidr* a virus spread … we saw ten [people die] in the night.
>
> (Older man, Barguna)

There also appears to be a strong link between the increased salinity and stillbirths:

> *In those regions where salinity has increased, abortions amongst women has increased – precisely because of high blood pressure – salinity increase*

> *and high blood pressure.* Salinity and high blood pressure have a relationship with abortion.
>
> (NGO Key informant, Dhaka)

Children's health is also particularly affected by changes in climate and weather events:

> Diarrhoea is one of the major problems for children … it is probably the biggest cause of high death rate amongst children.
>
> (NGO Key informant, Dhaka)

> The children are becoming sick with colds, cough, fever, skin diseases. [Look at my son-in-law.] The skin diseases spread on his hand, ear, hip, leg and many other places. This heat was not like now. Because of the heat – we are becoming affected by these diseases. Why are the children suffering from skin diseases? … I have pain inside my abdomen. Sometimes the pain becomes severe and my hands and legs stop [working]. I cannot even talk at that time.
>
> (Interview, older woman, Gaibandha)

Changing lives

In order to assess the gendered differences of these significant climate impacts, we asked women and men to tell us how women's and men's lives had changed as a result of the changing climatic conditions. Tables 7.9 and 7.10 indicate these responses. Table 7.9 outlines the perceived impacts on women and Table 7.10 on men.

The top four issues raised by all participants relating to women (and these are the four issues also prioritized by women) are:

- more health problems (61.5 per cent)
- increased pressure to acquire loans (48 per cent)

Table 7.9 How have women's lives changed in the last 10 years?

Changes in women's lives	*Women*	*Men*	*Total*
More health problems	56%	66%	61.5%
More stressed and anxious	46%	39%	43%
Increased pressure to provide food/shelter	12%	13%	13%
Increased pressure to acquire loans	47%	49%	48%
More pressure to migrate for work	16%	12%	14%
Better access to paid work	8%	8%	8%
More support from NGOs	38%	35%	36%
Experience more violence	13%	9%	11%

Table 7.10 How have men's lives changed in the last 10 years?

Changes in men's lives	*Women*	*Men*	*Total*
More health problems	47%	69%	58%
More stressed and anxious	43%	55%	49%
Increased pressure to provide food/shelter	28%	35%	34%
Increased pressure to acquire loans	18%	31%	25%
More pressure to migrate for work	21%	23%	22%
Better access to paid work	22%	15%	18%
More assistance from NGOs	12%	13%	13%
Experience more violence	8%	12%	10%

- more stressed and anxious (43 per cent)
- more support from NGOs (36 per cent)

By contrast, respondents were asked the same questions about how men's lives have changed in the last ten years.

The top four issues raised by all participants relating to men are:

- more health problems (58 per cent)
- more stressed and anxious (49.2 per cent)
- increased pressure to provide food/shelter (34 per cent)
- increased pressure to acquire loans (22 per cent)

These were the same priorities offered by men but with much greater emphasis. Thus men nominated:

- more health problems (69 per cent)
- more stressed and anxious (55 per cent)
- increased pressure to provide food/shelter (35 per cent)
- increased pressure to acquire loans (31 per cent)

Both women and men are experiencing, and are seen to be experiencing more health issues. Men are more likely to report that women have increased health problems, while women note that they are more stressed and anxious. Women are not under the same pressure to provide food and shelter but experience more pressure relating to acquiring loans, no doubt because of their access to microcredit schemes.

In subsequent questions relating to boys and girls it was noted that girls and boys are also perceived to be experiencing more health issues; girls are twice as likely as boys (170 and 81 respondents) to be under pressure to migrate for work. This suggests that girls are viewed as having an income-earning capacity in the garment factories.

Child marriage

Of note is that 45 per cent (276) of respondents reported that girls were now being forced into early marriages as a direct result of climate events and subsequent poverty:

> I heard from someone that a child marriage is going to take place in an area. Then, I went there and spoke to the family against child marriage. Then the marriage did not take place. The girl was not of marriageable age. She was only 12 or 14 years old. She was about to be married off at that tender age. Then we went there in a group and convinced her parents that marrying off their daughter at an early age will damage her health and will create a number of problems – the boy will demand dowry later, he will then beat the girl for dowry. The girl's parents understood that.
>
> (Focus group, female, Barguna)

Violence

Informants note that violence against women has increased as a direct result of disasters. Violence occurs in shelters, on the way to shelters, during and after a climate event and in the reconstruction phase. An extreme outcome is the rise of trafficking of girls and women from disaster sites to the capitals and across borders:

> In natural disasters, violence happens to women more, because the situation is not normal. Men aren't working, and tempers [are short] and they also suffer from various kinds of problems. The violent behaviour towards women within households and in the community becomes almost normal behaviour.
>
> (NGO Key informant, Dhaka)

> Then there has been this issue of violence. Women are now left alone [when men out-migrate], they're vulnerable. We're also worried about trafficking, but we can't pinpoint how many [are trafficked], because we haven't investigated enough. Children and women have been [trafficked] … In country or outside … From rural to urban. They can be brought in as domestic workers or in the brothels and sold off.
>
> (NGO Key informant, Dhaka)

Previous research from Bangladesh and indeed from across the world affirms that violence against women increases during and after natural disasters (see for example Alston 2013). One study examined this issue in a post-flood context in the regions of Faridpur and Gaibandha – two of the most flood-affected districts in Bangladesh (Nasreen 2008). All women in this study reported experiencing different types of violence during a flood, such as emotional, physical and sexual violence, and women with a disability were most vulnerable. Other studies highlight various reasons for increased violence against women following disasters. These include men venting their frustrations on women due to increased psychological stress and lack of employment, and punishment for not managing the increased workload or resources, not serving food on time and their inability to procure relief (Ahmed *et al.* 2009; APIT 2009; WEDO 2008). During and after natural disasters, women are at increased risk of sexual abuse and such abuse has been consistently reported in cyclone shelters (Oxfam and Naripad 2011; Ahmed *et al.* 2008; Nasreen 2008; WEDO 2008; Ahmed *et al.* 2007). Further, when husbands move away to seek employment, women and children lack protection in the home (Oxfam GB and Naripad 2011; Ahmed *et al.* 2008). Sexual abuse also occurs when walking to collect water (Ahmed *et al.* 2008; Nasreen 2008). APIT (2009) reports that increased sexual violence following Cyclone *Sidr* resulted in a rise in polygamy, early marriages and unwanted pregnancies, mental health disorders and sexually transmitted infections. Women's vulnerability to unwanted pregnancies is exacerbated by the interruption of supply of birth control pills following a disaster (Nasreen 2008).

Violence against women is widely accepted by the women and men who participated in this research. Many feel it is the right of men and part of the lot of women:

> Sometimes they slap us, once or twice if we have done any mistakes. We don't take it seriously. It's nothing. Suppose today my husband slapped me, so what! He is an old man who cannot control himself nowadays like the early times. I said something wrong that's why he slapped me. But you know after the incident he was sorry and consoled me.
>
> (Older woman Barguna)

> He beats me but he loves me also.
>
> (Young woman, Gaibandha)

More severe violence is less easy to justify or excuse:

> They [my husband's family] didn't give me sufficient food to eat. They used to give me food only once in a day. Beside this they did not provide me necessary oil and soap. They tortured me physically and still I have those scars on my body.
>
> (Young woman, Barguna)

An increase in violence against women is evident in the three regions examined and as a direct result of the climate changes being experienced. Most tolerate the abuse and excuse the behaviour as occurring under stress and as a normal, but exacerbated, expectation of living in rural Bangladesh with its traditional customs associated with women's disempowerment and men's privilege.

Out-migration

One of the most common outcomes of a climate event, whether catastrophic or slow-onset, is the consequent out-migration by a family member to source income to purchase basic necessities. As with most climate disaster sites, migration has become a common survival strategy for families who took part in this research. Often one or more members of a family will out-migrate for work and send money back to the village. Kartiki (2011) suggests that a more appropriate term is 'displacement', as many poor and landless people migrate through necessity and may not develop the resilience to deal with future climate stresses. One of the critical issues often overlooked when discussing out-migration is the class factor. When self-employed farmers, fishermen or business owners out-migrate they are often employed as labourers and wage workers. This impacts their class status as they move from middle to working class and has implications for self-perception and esteem.

Migration

The *push factors* for migration include:

- the destruction of livelihoods (farms submerged, crops destroyed);
- the destruction of houses, shelter and assets;
- landlessness;
- poverty;
- a lack of available cyclone shelters;
- unsatisfactory living conditions in shelters (including congestion, a lack of sanitation and a lack of privacy for women);
- insecurity for children (including the disruption of education, and a fear for children's safety if the family is living near embankments or flooded lands);
- a lack of optimism about any improvement in the villages in the near future;
- insufficient humanitarian relief; and
- threats to life as a result of potential disease risks.

(Kartiki 2011)

The *pull factors* for migration include:

- better pay and multiple work opportunities in urban areas;
- the presence of friends and families in other areas;
- prior knowledge of the destination;
- access to support systems in the destination community; and
- a lack of incentives to stay.

(Kartiki 2011)

The dominant form of out-migration amongst the participants is for men to migrate to the cities for work as rickshaw drivers or day labourers. Young women are also encouraged to go to work in the garment factories. Women are more constrained in relation to migration because of a lack of resources (including lack of information, financial and social capital, and social protection); family responsibilities and cultural factors make it difficult for women to migrate independently (Kartiki 2011). Nonetheless the growing trend for young women to migrate for work in the garment factories is challenging the historical reluctance or inability of women to migrate.

Out-migration results in i) less labour being available in the villages for farm activities; ii) young people leaving the villages; iii) a rise in female-headed households; and iv) an increase in polygamy or rural families being abandoned. Many express their disquiet at these issues. The increasing need for a family member to migrate for work has changed the nature of family relationships, and polygamous variations are emerging. Some men now keep two families – one in the city and one in their rural village. This creates hardship for women in the villages as their needs may be overlooked as time passes. This creates hardship for women who are virtually abandoned. Often their husbands will not divorce them as this would require that they hand back the dowry payment. Instead, they expect their first wives to be looked after by their families:

> My daughter's husband went to Dhaka and started rickshaw-pulling. Then after some days he opened a tea stall. He was having problem living alone (he needed someone, who can cook for him and do other household chores) so he sent for my daughter. My daughter's not interested. We told her that even if your husband keeps you under a tree, you have to be with him. So he took her. Within a year, he had run away with a young girl and left my daughter alone in the home. My daughter called me. Then my husband brought her back from Dhaka after hearing all these things. *Where would she stay?*, he thought. He brought her back to the village and sent my daughter to her in-laws. Now, we come to know that he (the husband of my daughter) has married another girl and moved back to his house with that girl. My daughter has stopped eating.
>
> (Older woman Gaibandha)

> Yes, … one of my nieces … Why was it necessary for her to marry a man who has another wife and children? She was in class ten. Is there any reason? … Why should she marry someone who has another wife? … He will rent another house to keep other wife … He says that he may not come that night. Giving food and other necessary things … Giving time to another wife he comes back to first wife. Same way he is dealing with the second wife.
>
> (Young woman Satkhira)

> They are marrying once, twice and the relationship is going on! What to do! We are tortured severely here by the men. They keep two wives at a time. He will not give you food but will feed the second wife. This way they keep two wives. Otherwise if he wants to give one a divorce, they have to compensate and give the amount of dowry (Den-mohor). He will not give divorce that will cost him. What they do, they force us and send us to our father's house. They say, for food go and work at others' house for food.
>
> (Middle aged woman Satkhira)

Many women discuss the out-migration of their husbands and the hardship they both endure as a result. Underlying these comments are the sense of uncertainty and vulnerability of women left in the village to run their households. As financial hardship escalates, there are also increasing expectations from men that further 'dowry'-type payments will be forthcoming from their wives' families to support the families in these new transitions. The increasing trend towards men abandoning their village families is causing significant hardship for women and their natal families:

> I am married and my husband used to do agricultural work. He goes to Dhaka and earns by pulling rickshaw and other jobs. [He does this] just to save our children. They are too young. They don't go to school. My brothers-in-law and their families live here in the neighbouring houses. And I live with my children. When my husband sends the money we spend it for food. We may move to … Dhaka to work in the garment factories and make some money so that we can start our life again. Then if we could buy land we will make a house. My husband said so. Then if I say how will I go to Dhaka, I have never been there! Then he says then you better bring 20,000 taka from your parental house.
>
> (Young woman, Gaibanda)

Out-migration is resulting in a significant rise in the numbers of female-headed households in the rural villages that are dependent on remittance income and responsible for doing any agricultural work. When husbands stop remitting, significant hardship results:

> This is a story of one particular hot spot, and the area is growing, where we have waterlogging and females are having miserable lives. I mean, this is the worst form of life and livelihoods there in waterlogged areas because we know places where the local people cannot use the land, the productive resource, for any production; nothing can now be produced there. A really terrible situation. … the males have all left their areas and more than 60 per cent of the households are led by women, not really led, they are burdened by the responsibilities, managed by women but under unacceptable circumstances.

> There are I think in the range of a hundred thousand families in those two subdistricts and 60 per cent of the families are managed by females because the males have all left, either temporarily or permanently. Now, these households, maybe 15 per cent or 20 per cent of these households, they don't remit anymore, they don't assume any responsibility whatsoever [for their village family].
>
> Men [are unable or] they're not ready to face that kind of [double] strike both from disaster and poverty; they leave their family. That's the worst kind of situation when a man just leaves the family because he wants to survive. But what about the women? We have seen proletarianization of women – female proletarianization … We say that a male proletariat has two hands to work, but a female proletariat has one hand, because in the other hand she is holding her child. No mother will leave children whatever disaster it is.
>
> (Workshop Dhaka)

The consequences of out-migration as an adaptation strategy and its growing significance as both a means of survival and a catalyst for abandonment are complex. While there is less work available in the villages as a result of climate disasters, crop failures and, in Satkhira, the trend to shrimp farming, the growing trend to out-migration has significant consequences for families and communities affected by climate disasters:

> [Climate change] has reduced the income opportunities for people. Previously a person with ten bigha of land could provide working opportunities for 10–15 people, but in a shrimp farm only one-fourth of that number can be employed. So people, both male and female, are migrating in search of work. [Some] people are migrating to India these days in search of livelihood. [Some] people are having to go into the jungle for work which is causing harm to their health. Previously when they used to cultivate and keep livestock for livelihood, they used to get fresh milk, fruit and vegetables to eat. But now that is not the case, they are also not getting proper treatment from the doctors.
>
> (Male, Focus discussion group, Satkhira)

One particular consequence is the loss of local labour as there are now fewer men available for day labouring in the local areas:

> In previous days they could hire men and net for catching fish but now we don't get men even by paying money. After *Sidr*, all the people have gone to Dhaka in search of work and somehow settled.
>
> (Male focus group Barguna)

> The crop is damaged; there are no men here, they migrated to other areas. They went to work as day labourers where there is work. Usually

> at this time people do not go [away]. But this year they migrated because there is no crop this year.
>
> (Older man, Satkhira)

Others, such as this mother from Gaibandha, comment on the out-migration of their children:

> [My son] started working [at Chittagong] as a labourer …. He digs soil but could not send any money until now. I called him yesterday. He cried out and said, this job is too hard. He cannot stand it. I sent one of my daughters to Gaibandha town. She is a maidservant. My eldest son also stays at Chittagong. He works at a brickfield. [My third son] does not go to school. He does not work yet. He goes here and there and tries to find food for himself only. [My younger daughter] I sent her to school. But she could not continue because of poverty. After I lost my home, I sent her to another family. But she could not carry the load. She was only 10 years old.
>
> (Older woman, Gaibandha)

Increasing work for women

The direct impact of climate change disasters on women is recognized by a number of key informants who have a more overarching view of the trends across the country. For women at village level, their experience appears a solitary, desperate one of neglect, shame, poverty and oppression. The wider trend, or the female proletarianization noted by one informant, is a widespread view amongst key informants. Female-headed households are very vulnerable, the responsibilities of looking after family resources are arduous and the destruction of food and water sources adds a significant dimension to women's work. Further, they may not have the resources to rebuild their homes:

> It's like the feminization of poverty, [this is the] feminization of the impacts of climate change – because women, to begin with, are dispossessed and marginalized. And then this [out-migration] leaves them even more vulnerable. Because men have to go and find livelihoods, or they're leaving. Women are still expected to remain where they are, and look after the children, look after – if they still have anything remaining – the cattle or food or crop, while somebody [usually their male partner] has to go out and get the food, cash or food and so on.
>
> Then the water sources have been destroyed, so it means women have to work longer hours to go and get water. Some of the areas I was working in, there was absolutely no drinking water. There was water, water everywhere – it's like, I couldn't believe how realistic that poem was. But none of that water can be drunk. So they had to walk miles. They

Table 7.11 Adaptation strategies most likely to be adopted

Adaptation Strategy	*Women*	*Men*	*Total*	*Comment*
Seek government support	54%	56%	55%	
Change main income-generating strategy	48%	58%	53%	More likely in Satkhira (64%)
Acquire loan to resume agricultural activities	51%	45%	48%	More in Satkhira (56%)
Acquire loan to buy food	38%	37%	38%	More in Satkhira (56%)
Seek NGO support	34%	27%	30%	30% overall but 41% in Gaibandha
Mortgage land	30%	23.5%	27%	More in Satkhira
Increase wage labour	21%	26%	24%	
Send children away for work	17%	22%	20%	More in Satkhira (25%) c.f. Gaibandha (19.5%) and Barguna (14%)
Take children out of school	20%	19%	19%	More in Satkhira (26%)
Borrowing	20%	15%	18%	More likely in Gaibandha
Seasonal migration	12%	17.5%	15%	More likely in Gaibandha and Satkhira
Sell productive assets	8%	14%	11.5%	More likely in Gaibandha
Spend less	9%	14%	11.5%	More likely in Gaibandha
Acquire loan for new business	11%	11%	11%	
Support from friends and relatives	9%	11%	9%	More likely in Gaibandha and Satkhira
Migrate permanently	7%	8%	8%	More likely in Gaibandha and Satkhira

would get one bottle or two. They don't have the cash, so they would buy a bottle or two maximum and then they would distribute it to the children while they themselves were drinking this rubbish water, getting diarrhoea. So, you would have to go through that process. Then, [if they are pregnant they are] delivering in such circumstances as flash flood, waterlogging. They have no facilities, but they are expecting, so they

> deliver in those circumstances. I don't want to paint such a horrible picture, but it is a reality [on the ground], and they're coping. Some of the places we work in, they are still living rough on the embankments because of this whole *Aila* tragedy.
>
> (NGO Key informant, Dhaka)

> Women suffer more for all reasons. One is, that men leave to go and get food, to find livelihoods, so in that way they are vulnerable and insecure. Then, when the family cannot manage, they might be taken out to another part of the country for work, and there's the risk of being trafficked, so their vulnerability goes up. Again in the coastal areas, there are issues. I've heard from women sex workers that, during this period, there is less cash available and therefore their customers are paying them even less and they are starving. And now the price of food has gone up so much.
>
> (NGO Key informant, Dhaka)

This research reveals the stark reality of the impacts of climate changes on the lives of those most affected – rural people dependent on agriculture and living in extreme poverty. What is also revealed by this research are the dire consequences for women and girls – their abandonment, their early marriages, their attempts to run female-headed households, to secure income for basic survival for themselves and their children, and the constant constraints that reduce their life chances. This research reveals that women and girls feel the very real impacts of climate changes more keenly. The overwhelming and most devastating consequences of climate changes fall squarely on the shoulders of women. Examining the adaptations adopted by women and men

Table 7.12 Negative factors affecting quality of life

Factors decreasing quality of life	*Women*	*Men*	*Total*
Lack of food	64%	43%	53%
Lack of drinking water	34%	29%	32%
Lack of paid work	50%	43%	46%
Access to paid work	22%	21%	21%
Poor health	29%	23%	26%
Access to health services	9%	3%	6%
Lack of adequate housing	9%	5%	7%
Reduced access to education	10%	6%	17%
Decrease in income	26%	39%	33%
Lack of support from NGO	6%	4%	5%
Lack of support from government	3%	6%	5%
Increase in expenditure	8%	29%	18%
Catastrophic expenditure	6%	9%	7%

Table 7.13 Happy memories in 2012

Top happy memory	*Women*	*Men*
Festivals	24%	15%
Family events	21%	10%
Good harvest	16%	7%
Business/work	1%	8%
Received loan	6%	0.5%

Table 7.14 What is worrying people

What are you worried about?	*Women*	*Men*
Domestic life	16%	14%
Financial income	12%	21%
Business	0%	10%
Employment/job	2%	5%
Children's education	25%	9%
Children's future	25%	8%
The future	7%	6%
Food	6%	5%
Health	6%	5%
Cultivation, crops, fishing, crabs	1%	6%
Daughter's marriage	5%	1%
House building/repair	1%	1%

shows both the limitations of women's circumstances and the possibilities of greater gender equality.

Reconstructing lives – adaptations

Because of the significant impacts of climate changes on people's lives, we examined how people adapt, how they might perceive the sustainability of these adaptations in the long term and what their views of the future might be. What emerges is not so much a critical understanding of the language of climate change – in fact such language is alien. Coping, adaptation, vulnerability and resilience mean nothing when one's children are suffering and there is not enough to eat. Rather, the overarching motivation for people experiencing the most critical challenges of the world's changing climate is the need to survive.

Previous research reinforces this prioritization – re-establishing secure livelihoods remains a priority for rural people living in poverty:

Table 7.15 Overall priorities for type of information people want

Type of information	*Women*	*Men*	*Total*
Alternative income generation	23%	27%	25%
Livestock management	27%	18%	22%
Microcredit	21%	20%	21%
Health	16%	20%	18%
Agricultural extension	10%	22%	16%

> We might want to talk about climate change; they want to talk about livelihoods. We might want to talk about adaptation for the future; they want to talk about changes that are happening right now, like those affecting their ability to farm'.
>
> (Raihan *et al.* 2010: 23)

Adaptation – alternative livelihoods

Nonetheless it is evident that survival means adopting various adaptive actions to survive. Thus rural Bangladeshis are adapting to climate change through such strategies as growing crops on higher ground (BRIDGE 2008) or growing crops that can be harvested before the flood season (Mitchell *et al.* 2007). The modification of traditional agricultural practices is the most usual adaptation implemented and includes changing rice varieties to enhance crop resilience, changing irrigation practices, promoting an integrated farming approach with changes in crop rotation and fertilizers, and innovative food production practices such as hanging vegetables and floating gardens.

Women adopt various practices such as rearing livestock, growing vegetables and fruit trees to contribute to household income (Raihan *et al.* 2010) and accessing microcredit to develop alternative livelihood strategies. In this and other research women express a strong desire for income-generating work, work that they can do close to home while attending to their various activities (Neelormi 2010). Previous research also reveals the extent of women's active involvement in adaptive strategies.

Table 7.16 Overall priorities for best ways to get information

Best ways of receiving information	*Women*	*Men*	*Total*
Telephone	16%	33%	25%
At market	11%	17.5%	14%
Radio	16%	11%	13%
Group meeting	13%	3%	8%
Community leader	9%	2.5%	6%

Box 7.1 Women's identified adaptation strategies

- crop diversification and adapted agricultural practices;
- support to learn about various aspects of agriculture and livestock such as flood- and drought-resistant crops;
- the suitability of different crop varieties (local and hybrid) for particular climatic conditions;
- the proper use of items such as manure, pesticides and irrigation, post-harvesting technologies and improved animal husbandry;
- relocating the community and/or family to an area within an embankment where they can live safely and cultivate land; and
- greater access to doctors, pharmacists, vets and agricultural extension services.

(Mitchell *et al.* 2007:14–16)

Additional strategies might include:

- accessing reliable sources of credit at low interest rates;
- seeking the provision of seeds, particularly vegetable seeds and tree seedlings, to grow vegetables;
- seeking small livestock such as chickens, ducks, and goats; and
- sourcing cash and food for work programmes which involve rehabilitation of local infrastructure used by, or of use to, the poor.

(Beck 2005)

In addition, the importance of the garment sector to national recovery strategies cannot be overstated. This industry employs many poor women who migrate from rural areas and provide much needed income for their families (Beck 2005).

Survey participants were asked about the strategies they adopt in the long-term reconstruction of their lives. Following a climate crisis, the adaptation strategies set out in Table 7.11 are those most likely to be adopted.

Respondents were asked to nominate whether there were, in fact, opportunities created by the weather changes. While most (65 per cent) suggested no opportunities, others noted:

- There is more day labour and increasing wages (45, only 8 of these in Gaibandha).
- There is increasing government and NGO assistance for food and shelter (67 of whom 36 live in Gaibandha). Women were more likely to nominate this opportunity.
- There is increasing government and NGO assistance for sustainable livelihoods (40).
- More groups and co-operatives are supported via external funding (27).

Negative factors affecting quality of life

In the light of climate challenges, participants in the survey were asked to comment on the factors that most constrained their quality of life. Many issues were raised, but the dominant ones reflecting very basic needs are not being met. Constraints mentioned include a lack of food, a lack of drinking water, a lack of paid work, a decrease in income and an increase in expenses. Table 7.12 lists these.

There are gendered factors associated with these constraints. Women are more likely to go without food in order to ensure their husbands and children have more. They will also restrict their intake of water if it means having to walk a long way to source fresh water and are more likely to note their own poor health. Men are more likely to note a decrease in income and an increase in expenditure as significant negative factors.

Attitudes

In order to assess well-being amongst our participants, we asked people to name their most happy memory of 2012 (Table 7.13). Forty-one per cent said there were no happy memories (140 men and 111 women). The following is a summary of the main activities that gave people a sense of well-being during 2012.

Participants were also asked to comment on what is especially worrying them. There were gendered responses to this question, with women more likely to note their children's education and future, domestic life, financial or income concerns and their future. Men were more likely to comment on financial circumstances, business or income and domestic life. (See Table 7.14.)

Other issues mentioned include the political situation, social conditions, pirates, their own wedding, rising salinity, the next flood, religious sustainability and a daughter whose husband has rejected her.

Our survey participants were asked what is the most important type of information they required to re-establish their lives following a disaster. There are gendered differences in the type of information people nominate (see Table 7.15). The priorities for men are information on income generation, microcredit, health care and livestock management, and the priorities for women are similar but with slightly different focus. They suggest information on livestock management, income generation, microcredit and health will assist them to reconstruct their lives. Noteworthy is that women and men want information and assistance to re-establish their livelihoods into the future.

Critically it is important to note that women and men do not receive information through the same processes and that transmitting information to people at village level remains a challenge. It is difficult to assess the best ways to deliver information to people in the villages, as there are few evident sources that successfully reach large numbers of people. There are, however, gendered differences in the nominated options for receiving information, a factor that must be acknowledged in order to ensure that women have access to early warnings and ongoing information to assist adaptations (see Table

7.15). The preferred sources nominated by men in our survey are telephone, sharing information at the market and the radio. The preferred sources for women are telephone, radio and at group meetings. Other sources specified were newspaper (15), relatives (14) and the local school teacher (12). Overall it appears that women have less access to all forms of communication, a factor that should be addressed in policy and practices relating to disasters.

However, it is also important to bear in mind other factors that shape and restrict women's lives. Particularly important is women's low involvement in decision-making. Women are making limited decisions in almost all areas, a factor that severely restricts women's capacity to adapt effectively. In the interests of protecting and saving lives, addressing women's disempowered position is essential.

Moving beyond disempowerment

Nonetheless, our research makes clear that, in the aftermath of disasters, women are moving beyond the home, drawn by the necessity to earn income. The movement of women into the paid workforce discussed in Chapter 4 is accelerating with climate changes, as the need for income becomes so much more critical. Women are streaming into paid work, work that brings its own income and allows a degree of autonomy. There are growing numbers of women now living in female-headed households either because the disaster has taken their husband's life or because he has moved away to earn income for the family. These major social changes are having a significant impact on the freedom of women to move beyond their households. It is this critical space created by climate challenges, born through hardship, and manifest in critical need, that may provide a more empowering space for women. Women who took part in interviews and focus groups noted these changes during their lifetimes – from being veiled and silenced to now being free to move beyond the home. Now women are working in the fields and in paid work and go to the markets on their own:

> After *Sidr*, the number [of women working outside] has been high. Not many women worked on the roads before, say 10 to 20 was the number in a Union who worked outside. But after *Sidr*, 80 per cent of women now work outside.
>
> (Older man, Barguna)

> Women are moving into non-traditional jobs as social structures and the social safety net once provided by extended family, neighbours and community are breaking down.
>
> We do have examples from South Bangladesh where women, in the absence of their husbands who perished in the typhoon, have taken up fishing as a livelihood opportunity. It's not a one-solution-fits-all sort of a thing. It's very localized. It also depends on the social safety net, which is breaking down now. We always felt there was a social safety

> net – extended family, neighbours, etcetera. I think the repeated battering of the coastal areas has diminished that social safety net factor.
>
> (NGO Key informant, Dhaka)

However, there are also instances where the old meets the new and where non-traditional work meets customary law. In these instances women are making adaptations, or perhaps maladaptations, that appear harsh. In one village where many fishermen were lost in the cyclone, women are trying to fish without being seen:

> [In one area] at 4 o'clock in the morning, so it is still dark, I saw women using fishing nets to collect what is less than half a dollar, 45 cents worth of money they could make from 4 to dawn. Why this time? Because no men can see them fishing. … So, they wake up and they fish on makeshift sort of bamboo platforms, where, inside the dwellings there is already water [and where they try to remain hidden].

However as women themselves explain, necessity often overcomes customs and they must work to survive:

> We are needy; we need something to do to earn. If we do the job covering our body and head according to the custom of Islam will it do any harm? Poor Muslim women need to eat also, isn't it? We work hard to earn.
>
> (Female focus group, Barguna)

Men as well as women are acknowledging this necessity, and the accompanying changes in gender relations, and are more accepting of women working in public:

> Now when a girl works in an NGO after completing [schooling] I appreciate her. Older people have also changed their perspective. A few days ago an NGO worker (a woman) drove a motorbike. Men were looking at her. Even a girl said 'look – women ride bikes nowadays!'
>
> (Older man, Barguna)

Changes are also evident in the services and support available to women, including family planning and birthing. These are often the result of successful NGO campaigns, but the end result is that women have access to services that may save their lives and this certainly gives them added freedom.

> When I got married, I did not know anything about that. But girls now know many things. They do not need to take advice from others. They know what to do before and after menstruation and what should they take.
>
> (24-year-old woman, Satkhira)

> Now almost everybody goes to hospital. There are few cases [of childbirth] that is done at home. If normal delivery happens we do not take her to the hospital. We observe 6 hours as we are advised, and after 6 hours we take her to hospital.
>
> (Older woman, Satkhira)

Backlash

However, as many people were quick to tell us, there is also a strong backlash against women's growing freedom, and actions are being introduced to restrict women's rights. These are evident from individual, to community, to national levels. Some place the blame for climate challenges on women themselves and hold them responsible for the problems in society. The backlash has resulted in a rise in early marriages, a strategy to curb women's power, and attempts to restrict the rights of girls to education. There is also criticism that women and girls are taking employment away from men and boys:

> Women were riding bikes and accessing loans from the [microcredit agencies], and all these organizations brought in a revolution. It's not happening now. We find the younger girls getting married. Girls who have managed to get an education don't want to get married. So, there's no one description; it's become very complex … Our generation was out there all over the place, but this generation, we are not finding the same.
>
> (NGO Key informant, Dhaka)

> Nowadays women fight for equal rights. Girls tell that we can do everything. Beside this the job rate of girls has been increased so that a boy loses the opportunity to get that job. So the dictatorship of women has been increased. The education rate of girls has been increased after 1971 and the restriction of wearing religion dress has been reduced.
>
> (Older man, Barguna)

> We have heard this at a religious meeting. They say we do not listen to our husbands, we do the wrong thing and for this the weather is changing. This imam does not listen to the news. So he does not know how the weather actually changed. We cannot get them to understand that the weather is changing because of smoke from mills and factories. Their opinion is that for this disaster we women are responsible.
>
> (Young woman, Barguna)

> Unless you address the fundamental issue of women's right to property resources, the status of women will remain as second-class citizens.
>
> (NGO key informant, Dhaka)

Summary

At individual and community levels, coping and adaptation are fraught with the stark reality that mere survival is often the only coherent strategy. Villagers are constrained by their basic requirements for food and water, livelihood development and safe sanitation. Livelihoods are immediately affected after a disaster, health is compromised, women and children are malnourished, crops and productive land are devastated, and cropping cycles and traditional knowledge are no longer relevant to the precarious circumstances in which many find themselves. An obvious consequence is the out-migration, usually of men, to source income. This has short-term gains for the family but over time is compromised by emerging practices of polygamy and abandonment, early marriage and overwhelming poverty, and increased violence and abuse. For women and girls the realities of life during and after a disaster can be extreme and are affected by cultural practices, a lack of safety and security, and compromised lifestyles.

Adaptations in the aftermath of a climate event require significant changes to people's lives and livelihoods. Often this includes women moving beyond the home to undertake paid work, to work in agricultural production or to increase their own livestock management or other enterprises. This movement of women into paid work has opened up opportunities for a new way of formulating gender relations and new possibilities for gender equality in the post-disaster space. It is also clear that there is a backlash against women that may reconfigure inequitable gender relations. NGOs are very aware of the need for gender equality and rights-based practice. In the light of the development of new Sustainable Development Goals post-2015, there is global attention to principles that should guide global development. It is clear from our work that gender equality should be at the forefront of these goals if we are to progress the rights and circumstances of the most critically affected and vulnerable people on the planet:

> There has been an attempt at trying to look at how policies are climate sensitive and whether gender issues are adequately reflected, practically little. The problem is, everybody understands at a conceptual level that this is necessary, but how does one translate that in terms of actual policy or action or implementation? That is not yet clear.
>
> (NGO Key informant, Dhaka)

Notes

1 A bigha is a measure of land from 1/3 to 1 acre.
2 A shatak is equivalent to approximately 450 square feet.

8 Conclusion – vulnerability, resilience and adaptation

Advancing gender equality

This chapter brings together this research on climate change in Bangladesh with a particular focus on the impacts on women. It outlines findings on climate change, vulnerability, coping, resilience and the types of adaptations adopted to improve and enhance quality of life and gender equality in this highly vulnerable country. As such it is globally relevant – the insights drawn from this work and the ideas for strategies and actions to empower women and girls are applicable in other contexts and sites.

Bangladesh is a relatively new country with an already rich history and a vibrant culture. It is a country struggling with significant vulnerability to climate change as it metamorphoses from a highly vulnerable developing world status. Critical and far-sighted policies have been enacted to address poverty, to improve livelihoods and infrastructure and to educate its children. These developments are threatened by the climate challenges facing the country.

This book has outlined its history, culture, laws and policies, and noted the climate threats facing the country. What it also does is bring forth the voices of the people most critically affected, giving shape and substance to lives often overlooked in the global hothouse of environmental politics. By advancing the voices of Bangladeshi people and describing their lives, their wishes for their children, their wishes for survival, work and safety, this book challenges the global community to acknowledge the impacts of climate change on global citizens and to adopt practices and policies that give the women and men, girls and boys of Bangladesh the opportunity for an equitable future. Of particular concern to the writer is that the impacts on women and girls be addressed and that climate change and post-disaster scenarios not be used as sites where gender inequalities become entrenched. Rather it is in these spaces where new forms of gender relations can be enacted through education, work, safety and exposure of the potential for a backlash against women.

Gender equality in the context of climate change

Bangladesh is an Islamic country where the principle of gender equality is embedded in the Constitution as a result of the historical struggle with

Pakistan in the 1970s. The engagement of women as freedom fighters and their experiences during the war led to a strong validation of women in the immediate aftermath of the country's independence. Over the subsequent four decades the recognition of women's rights has advanced and retracted with various changes of government. Further, there is a constant tension between constitutional and customary law that results in constraints on the freedom of women and girls. These tensions often turn violent when fundamentalists challenge gains made by women. The fight for gender equality is an ongoing and deeply ideological battle.

It is against this background that we must view the challenges posed by climate change to the people of Bangladesh and subsequently the impacts for women and girls. Bangladesh is one of the most vulnerable countries in the world, experiencing a range of catastrophic events with increasing frequency, and these have not only had a significant impact on the social, environmental and economic stability of the country but have also led to critical changes in the lives and livelihoods of Bangladeshis. Climate change adds a new and volatile uncertainty to the people and communities that formed part of this research.

Bangladesh has been extraordinarily successful in reducing the loss of life from climate disasters and in continuing to reduce poverty and famine. The scale of the disasters, the increasing frequency of events and the additional factors associated with interrupted crop cycles, reduced food production, uncertain weather, a high population base and major obstacles in infrastructure associated with transport and communication has meant that this battle is overwhelmingly difficult.

There is no doubt that the Bangladeshi government together with international and national NGOs are building and developing adaptive strategies to save lives and to enhance the quality of life for those living in some of the worst affected areas on the planet. At government levels this includes advanced planning systems, protocols and practices to predict weather patterns and to give advanced warnings to communities. At NGO levels it includes assisting with immediate crisis interventions, reconstruction efforts and long-term adaptive strategies designed to improve the lives of Bangladeshis. We have witnessed very practical and life-changing solutions such as the building of new ponds, water-filtering devices, food security measures and training activities, all of which are moving Bangladesh forward in the quest to deal with climate changes – an evolving and highly destructive phenomenon. More intangible is the pursuit of human rights for all, including women and girls. These are evident in the programmes and policies of many NGOs – a process that is normalizing the expectation that women and girls have equal rights to a secure future and that they will be a critical part of ongoing activities.

However there are many issues that constrain and reduce these efforts including governance structures, the capacity to deal with the scale and frequency of disasters, the resource limitations that make it difficult to reach all who are impacted by disasters and the difficulties associated with ensuring that all receive early warnings, aid, reconstruction assistance and social supports. The ongoing tension between constitutional and customary law in the context of climate disasters is reshaping gender equalities through greater freedom of movement for women in the context of livelihood strategies and the need to earn income. On the other hand there is evidence of a backlash against women that is most obvious in early marriage, the acceptance of violence, dowry disputes, the inequitable distribution of resources and a failure to incorporate women equitably in decision-making forums and capacity-building exercises.

Ferree's (2012) notion of *radical realism* allows an analysis of the possibilities that climate change provides for a reimagined space where gender equality is centralized. She argues that ideas that may appear radical in a particular society at a particular time may appear quite tame at others. Thus the idea that women in Bangladesh should have freedom of movement in order to develop their livelihood strategies appears tame in the context of climate disasters and yet is a radical departure from that of previous generations. We would extend this further, arguing that a new concept, *radical adaptation*, should be a part of climate change terminology. This concept describes adaptation policies and practices that encompass gender equality and human rights as taken for granted and transparent in any actions or strategies designed to assist communities. Radical adaptation is necessary if the global community is to effectively address the impacts of climate change in positive, practical and just ways.

Our work has demonstrated that women and girls have in many cases been catalyzed by climate change. They are significant players in the battle to preserve and maintain social structures, they are moving into paid work and are active in their fight to protect and safeguard their families. Yet there is also increasing evidence that women and girls are vulnerable in climate change spaces. They do not own the resources necessary to rebuild lives post-disaster, they are largely responsible for their families, they put themselves last in relation to food and water access, they are increasingly heading households as men and boys migrate away to secure income, they struggle with customs that demean them, they are experiencing increasing violence, early marriage, dowry disputes, trafficking, food and water insecurity, increased care tasks and are responsible for the health of themselves and their families. They are the most likely to die in disasters and are particularly vulnerable in disaster and climate change spaces.

Nonetheless this research reveals that women have a strong family identity; a unique knowledge of their communities, landscape and agricultural techniques such as seed-saving and livestock-rearing; a powerful spirit of cooperation; a thirst for education for their daughters and sons; a vibrant culture and an indomitable spirit. Women and girls are critical agents of change in circumstances where climate change is creating major challenges and are critical to the global battle to preserve our planet and its diverse cultures.

By taking a gendered approach to this work on climate change impacts we demonstrate that the gender order of a society – the level of laws, customs and practices, which shape unequal futures for women and men – constrains individual women in their gender negotiations at the intimate level of their families and communities. Women are bought and sold in marriage through dowry payments and other practices that diminish their capacity and limit their education and opportunities. They have little independent right to resources and must address the daily challenges of food and water security in their 'everyday/every night worlds' (Smith 1992). They are at greater risk of becoming the collateral victims that Bauman (2011) warned against: 'the "collateral victim" ... of any natural catastrophe is currently one of the most salient and striking dimensions of social inequality'. (Bauman 2011: 7)

Thus we agree with Enarson's (2012) appraisal that gender inequalities, environmental degradation and vulnerability to disaster are so strongly linked that ignoring gender risks perpetuating degradation and reducing our capacity for adaptation. Women through their *everyday/every night* work, are central to adaptation and sustainable practices. A failure to recognize this risks endorsing non-sustainable coping practices. Women will continue to eat less, they will continue to school their children to drink less, they will continue to find short-term ways of coping in order to survive to the next day and the next week. These practices are unsustainable in the long term and will lead to major health crises, to children dropping out of school and to other practices that will exacerbate wealth differentials between rich and poor. Moving people from unsustainable coping strategies to long-term adaptation practices oriented to livelihood security, the efficient use of resources and long-term planning requires attention and resourcing from global, national and local institutions.

Gender is critical to this planning and action. Without gender mainstreaming we risk cementing gender inequalities in policies and practices, and treating women and girls as the collateral damage of global actions. It is not enough to argue that cultural relativity excuses those well-intentioned institutions and agencies that shore up women's degraded space on the planet. This is not what women want. They want the dignity of their lives to continue, to have the capacity to make their own life choices, to marry when they are ready to marry, to undertake livelihood strategies using their own resources, to make decisions about their lives, to have freedom of movement, to have the capacity to protect their children, to nourish and nurture their families, communities and landscape.

Adaptation

Addressing adaptation strategies therefore requires attention to the social and gendered elements of a society. Moser and Ekstrom (2010: 22026) define adaptation as 'involv[ing] changes in social-ecological systems in response to actual and expected impacts of climate change [environmental factors] in the context of interacting non-climatic changes'. This definition makes more explicit the non-climate factors that frame the capacity of people to adapt in socially desirable, equitable ways.

Elsewhere (Alston 2013) I have argued that social systems include inequality, access to technology, power relations, conflicts, values, knowledge, policies, cultural context, governance and institutions, all of which have gendered elements that determine levels of vulnerability. The opportunities for, and pressures on, people and ecosystems affected by environmental factors are heavily dependent on the interlinked social and ecological systems that have generated them. To ignore social systems risks undermining positive adaptive responses, enhancing existing inequalities and increasing the vulnerability of women.

However, it is arguable that the current limited focus of adaptation research on scientific and technological solutions masks social inequalities and inherent vulnerabilities. Much adaptation research is based on the fundamentally flawed assumption that adaptation is both possible and desirable in all parts of the globe and that, with determined effort, the right technology and adequate funding, the vulnerable will adapt. This linear assessment of adaptive responses presupposes that existing conceptual frameworks are adequate. Yet it is increasingly clear that there are limits and barriers to the capacity of individuals and communities to adapt, and that expectations that rational adaptive responses will occur are illusory for a variety of reasons, not least being widespread poverty and inequalities amongst those most vulnerable (Tschakert *et al.* 2011).

Resilience

Successful adaptation relies on the resilience of people and communities. The IPCC (2007) defines resilience as the capacity to absorb change in a positive way. To move people from coping to adaptation is highly dependent on how resilient people feel, how risky change appears, how safe traditional practices seem and the institutional supports available to assist people to move through uncertain change. Barriers to resilience and successful adaptation include the *physical* aspects of the disaster, *financial resources* (levels of poverty and the costs of adaptation), *cognitive aspects* including the assessment of risk and trust in institutions, *normative behaviour patterns* including continuing what has always been done and what appears safe, and *institutional governance and structures* (Alston 2013; Jones 2010; Koelle and Shackleton 2011; Pelling 2011; Adger *et al.* 2007 and Tschakert 2011).

In the regions that form part of this research these barriers are very real. The *physical* aspects of the disasters are intense, highly destructive and frequent – eroding the resilience of people and communities with their savagery. The *financial resources* are limited and require the sourcing of alternative income, and subsequent out-migration, as a common form of adaptation. The costs of adapting in situ are undermined by the intensity of the events experienced, the ongoing damage, the major reconstruction required and the fear of further events. *Normative behaviour patterns* may continue because of a lack of alternatives and few resources. However, for many people in this research, continuing to do things as they always have is often not possible. In Sakhira for example the salt water inundation of once fertile lands prevents cultivation and has led to the development of shrimp farms in the saline water. Arguably this new industry is unsustainable or may cause further problems in the long term.

Trust in *institutional governance structures* varies. There are high levels of trust in NGOs and local organizations but more limited trust in layers of government that operate beyond the local. The lack of power and influence that local people feel they have on national policies and international actions leaves many feeling powerless to influence reconstruction efforts, resource distribution or livelihood reestablishment. There would seem to be an evident need for gender-equitable, village-level decision-making bodies to be linked to union level and beyond in representative, transparent ways so that the views of local people can reach and influence those at the highest levels of institutional structures.

Building resilience is dependent on these factors and is essential to successful adaptation. Resilience is also essential to gender equality and is defined by women's empowerment in critical circumstances. If the global community is to foster and develop transformative adaptive change (Pelling 2011), resilience based on gender equality and the empowerment of women is essential. This requires governments and institutions to redefine their social contracts in a similar way to some of the NGOs quoted in this book. The transformation of the work of these NGOs from aid distribution instrumentalities to advocates of human rights and women's empowerment is now widely accepted in communities. A similar approach by the Bangladeshi government, based on their constitutional law and the revised women's policy, is a good place to begin. To do this requires challenging the fundamentalist viewpoint relating to women as one that is counterproductive to national progress, development and the very survival of the country. Overt actions must be taken to address the rights of women and girls to safety, security, equality and the right to make their own decisions about education and marriage.

Climate change – bringing forward the voices of the marginalized

Having spent time in this amazingly beautiful country, I find it evident that Bangladesh stands at the forefront of climate change impacts and yet must watch as the developed world argues about responsibility, causes and climate taxes. The release of the fifth assessment report of the IPCC confirming that there is now 95 per cent certainty that humanity is responsible for climate change (IPCC 2013) signals a time to stop procrastinating, to move forward with radical solutions, to shift the focus from the science to the human impact, from dithering to action. The confluence of the fifth report with the development of the Sustainable Development Goals for 2015 and beyond provides a timely coming together of science and justice. Debate on the Sustainable Development Goals is focusing the world's attention on how we might move forward with principled social justice initiatives, including gender justice.

With the increasing frequency and intensity of climate events, the way that climate change is represented and interpreted demonstrates who has the power to define and shape strategies and responses. Overwhelmingly it has been represented as an economic and environmental problem, with much more limited attention to social impacts. Unspoken principles underlying much of the discussions of climate change are about productivity and resource security. It is a depopulated space where social and gender justice rarely feature. This book brings forth the voices of the people of Bangladesh to challenge that framing of this wicked problem. Climate change is about people, it is about the challenges to lifestyles and livelihoods, to health and safety, to dignity and respect. Until we understand these components of climate change and the impacts on the poorest of the poor and the least able to withstand and adapt, we will not find adequate solutions.

To do this, global and national actions must incorporate gender mainstreaming in all policies and actions and global strategies. This does not mean adding women in and stirring. It means understanding the world from the view of women and prioritizing gender equality, in order that resources are distributed evenly, that the work and contributions of women are valued and that their lives are enhanced and protected.

Post-disaster – space for a global commitment to gender equality

The post-disaster space provides a new arena to reconstruct not only the physical landscape and infrastructure, but also unequal gender relations that hold back national reconstruction. Unless gender relations are reformed as part of reconstruction efforts, strategies and actions aimed at adaptation risk reinforcing and cementing gender inequalities through an unquestioning acceptance of rights abuses. If a gendered division of labour is normalized in climate change and post-disaster responses; if gender-based violence is accepted as traditional, and therefore unchangeable (Zalewski 2010); and if women are viewed as essentially mothers and carers and men as landowners, fishermen and workers, then interventions can lead to increased gender-based violence, inequitable resource distribution and a lack of commitment to empowerment strategies for women. However, if all are viewed as equal, then these gendered assumptions cannot be reinforced. A reframing of climate change policy and actions, and gender-sensitivity to women's circumstances, has the capacity to move people and communities to transformative change and to shift and challenge inequitable gender relations.

When women's rights, responsibilities and opportunities are recognized; when they are given equal influence on national priorities; when they are an equal part of decision-making on resource allocation and policy; when they are integrated meaningfully into the workforce with equal and living wages, major advances can be achieved. When women are empowered, there are better national health and education outcomes and a more sustainable use of natural resources. Gender-sensitive national and international institutional frameworks and actions are required to assist vulnerable countries and people within countries to adapt. The UN Secretary General Ban Ki-Moon notes the critical need for women and girls to have an equal role in climate change decision-making and to be recognized as critical agents of change.

Benefits of gender equality

Perhaps not surprisingly research is emerging that indicates that gender sensitivity in post-disaster situations has very positive social and economic benefits for women and men. International research suggests that women exhibit greater awareness of, and responses to, environmental problems (Goldsmith *et al.* 2012; McCright 2010; World Bank 2009) and that there are gender differences in environmental attitudes, concern for the environment, knowledge levels regarding climate change and willingness to take action to improve environmental problems (Bord and O'Connor 1997; Semenza *et al.* 2011; Zelezny *et al.* 2000: 159–160).

International research also confirms that the engagement of women in post-disaster restoration significantly increases positive outcomes. The UN International Strategy for Disaster Reduction (2007: 9) notes that 'attention and resources directed to local, women's … traditional cultural knowledge and expertise produced more efficient, relevant and cost-effective projects'.

As documented in previous work (Alston 2013d) the UN International Strategy for Disaster Reduction (2007) notes that gender sensitivity produces significant environmental, economic and social benefits, including:

- more land now under cultivation by small farmers, healthier soil, better production, more crop surplus and resultant income, increased self-sufficiency, more and better protected green areas, growth of culturally important plants and increased involvement of young people;
- communities better organized and prepared, and more aware of impending disasters;
- information disseminated more effectively, community members more readily assisting one another, evaluating damage, monitoring reconstruction, fairly distributing resources and interacting with donors without being manipulated;
- the economically most vulnerable, subsistence-driven farmers – most of them women – becoming more involved in decision-making, planning, implementing and monitoring;

- poverty being better managed through sustainable agriculture; and
- subsistence farmers managing the environment in a way that reduces risk, monitoring disaster-resistant housing construction and becoming surveyors of risk and damage, trainers, political advocates and leaders.

Additionally they note that involving women in disaster-risk reduction and reconstruction has unexpected results, including that housing design better meets local needs, that projects have been integrated into schools, that green areas are protected and that resources are more fairly distributed (UN International Strategy for Disaster Reduction 2007).

A global women's movement

In instituting feminist values into climate change actions and practices and addressing gender inequalities in fragile states, the need for a global women's movement becomes evident. As Pakulski (2004) so cogently argues, feminist movements in very traditional societies are weak because of the strong legitimization of widespread discriminatory practices by religious beliefs and customary laws. Nonetheless, women's capacity for organizing around issues such as environment, education and health are evident in many countries including Bangladesh (see for example Agarwal 2009). While a global women's movement will assist to empower women in diverse cultural contexts, we must ensure that differently positioned women do not presuppose a Western notion of superiority, that an intersectional understanding is adopted and that scholarship on women's activism is decentred beyond Eurocentric and Western experiences (de Haan *et al.* 2013: 5–6). Intersectionality is a critical aspect of ongoing feminist activism. As Ferree (2006: 10) notes 'no one has a gender but not a race, a nationality but not a gender, an education but not an age. The location of people … is inherently multiple'.

Thus, we argue for a global feminist movement to acknowledge the impacts of climate change on women and girls in diverse contexts, to centre these experiences in emerging knowledge and to advocate for women's rights to resources, aid, local organizing and representation, particularly for women in fragile states and vulnerable communities with few resources and high levels of poverty. In particular we note the need for an appraisal of the politics of gender within global, national and local discussions, forums, debates and policies in order that these prioritize the vulnerability and empowerment of women and girls.

A global movement is achievable as most countries now have women's machinery, and many transnational organizations have gender equality as a goal. Feminist scholars across the world are also developing new knowledge on old problems and reimagining new ways of enhancing equality. A global movement could link these groups, forging a web of strategic alliances with transnational organizations such as UNWomen, national women's ministries, other civil society organizations, local collectivities of women, academics, activists and gender experts within governments. Snyder (2006: 24) terms the UN the *unlikely godmother* of a global women's movement and views this as an ideal leader in the formation of such a movement.

In Bangladesh there is an evident need to link local women's organizing with national-level policymaking and transnational activism. Without these links, women at local rural levels remain locked into disempowered positions, subject to significant pressures that are magnified in the context of climate events.

A global movement has the potential to expose the taken-for-granted gender relations in local customs and in aid distribution, resource allocation, decision-making and prioritizing of actions. This will lead to a focus on adaptation and women's resilience, and a decline in women's vulnerability in post-disaster spaces.

We leave the reader with a challenge – to recognize and act on the rights of women who are living in one of the most vulnerable countries in the world to climate change. Responsibility for incorporating women cannot be left to institutional bodies that have too often ignored women. There is a demonstrated need for a vibrant and activist international women's movement to advocate on behalf of all women and to provide support to national women's movements (de Haan *et al.* 2013). In Bangladesh there are strong women advocates and a historically vibrant women's movement. There is much to be done in national and international spheres to ensure that women in Bangladeshi villages are empowered to fight for their rights and to be recognized for the critical role they play on the front line of climate disasters. Perhaps it is time for a new conversation on climate change – one that is populated by women and men, girls and boys – people at the forefront of the world's climate change actions, people amongst the poorest of the poor. As the world debates the Sustainable Development Goals, there is a new urgency to this discussion. Including gender equality as a fundamental global goal for the future is more than an intellectual debate – it is a critical factor in the future of the planet.

Conclusion

Our research indicates the greater vulnerability of women in areas affected by climate-induced disasters and therefore the centrality of gender in any policies and actions designed to address adaptation and build resilience. We have noted that many rural places where environmental and climate events are located are already undergoing significant restructuring and transformation. The addition of climate-induced disasters adds to already complex processes of adaptation.

We argue that the language of climate change is constructed around a scientific reductionist view, centred on productivity and neoliberal ideas of market supremacy, thereby reducing attention to human rights, social implications and gender inequalities. We argue that gender is a significant factor regardless of cultural contexts and that embodied gendered practices are evident in the representation and responses of women and men. We note a global gender order that is facilitated by nation states and cultural practices, resulting in significant disadvantage to women.

In post-disaster sites, unless we can 'see' the implications of gender equality to human well-being, reconstruction efforts can significantly disadvantage women. Not only are women fighting poverty, they may also see control of their own resources accorded to men through the misguided assumptions of patriarchy. Post-disaster sites provide room to build gender equality measures into actions and activism.

Nonetheless, the benefits of gender equality measures are so significant that ignoring gender in policies and actions risks significant social disintegration and *collateral damage* on an unprecedented scale. We present a conceptual framework, drawing on feminist political economy and ecology to argue that the link between gender, disasters and sustainability is irrefutable. We therefore urge that women at local levels be supported by the state and by a global women's movement that will bring the voices of marginalized women to climate change discussions, policies and actions.

Appendix

Policy and action

National policy framework

Through international protocols such as the Hyogo Framework for Action, United Nations member countries including Bangladesh have committed to climate change actions. In addressing these commitments, Bangladesh has adopted a climate mainstreaming approach involving several ministries in actions to address climate challenges. As a member nation, Bangladesh has also developed its UN-mandated National Adaptation Programme of Action (NAPA), a document that spells out climate change adaptation measures (GoB 2005). Subsequently the government has also developed a Bangladesh Climate Change Strategy and Action Plan (BCCSAP) detailing strategic actions for climate change mitigation and adaptation (GoB 2009). The NAPA considers urgent and immediate priorities for adaptation, while the BCCSAP is more comprehensive, with medium- and long-term actions on emissions reduction, adaptation, mitigation, technology development and transfer, and financing (Alam *et al.* and 2011).

Box A.1 The National Adaption Programme of Action adaption strategies

The National Adaptation Programme of Action outlines interventions and adaptation strategies based on existing coping practices and suggested future strategies. Importantly, it prioritizes:

1 the impact of climate change on the lives and livelihoods of communities;
2 poverty reduction and sustainable income generation;
3 enhancement of adaptive capacity in terms of skills and capabilities at community and national levels;
4 gender equality (as a cross-cutting criteria);
5 enhancement of environmental sustainability;
6 complementarity with national and sectoral plans and programmes; and
7 cost effectiveness.

Despite gender equality being prioritized in the NAPA, it does not include women as stakeholders in proposed adaptation actions, does not incorporate gendered statistical information and does not discuss the gendered implications of public health issues. Rather, 'the NAPA presents women as victims of climate impacts; women are not considered active participants in adaptation to these impacts' (WEDO 2008: 59).

Further, concerns have been raised about the capacity of the NAPA 'to fully consider local knowledge and local priorities regarding climate variability and change' (Thomalla *et al.* 2005: 13), as this local knowledge is considered by many to be critical to positive adaptation (Raihan *et al.* 2010; Jennings and McGrath 2009). Local-level actions, empowerment strategies, capacity building and less centralized authority are viewed as essential (Agrawal *et al.* 2009: 15):

> Many local politicians and government officials report that they have not received any policies, directions or funds for climate change adaptation, and they are therefore reluctant to do anything.
>
> (Raihan *et al.* 2010: 58)

The second, longer-term document, the Bangladesh Climate Change Action Plan (BCCSAP) (GoB 2009), is a 10-year plan to build Bangladesh's capacity and resilience to deal with climate change.

The Bangladesh Climate Change Action Plan outlines six pillars of action including:

1 food security, social protection and health;
2 comprehensive disaster management;
3 infrastructure;
4 research and knowledge management;
5 mitigation and low carbon development; and
6 capacity-building and institutional strengthening.

The BCCSAP is the officially endorsed government strategy regarding climate change; yet it also lacks attention to local knowledge and has limited participation from women, vulnerable peoples and NGOs (Raihan *et al.* 2010; Alam *et al.* 2011; Hossain 2009).

Both the NAPA and BCCSAP focus on natural disasters and give limited attention to incremental climate change. Nonetheless, it is evident that there has been little progress in implementing the goals and targets of the plans (WEDO 2008); that the scale of implementation for these plans is large; and that the resources may not be adequate. For example, with the predicted population growth, by 2050 more than 9 million people will need emergency shelter during storm surges, requiring an estimated 5,702 new shelters (Dasgupta *et al.* 2010). Thus, while Bangladesh may be actively responding to climate change through policy development, its on-the-ground action and capacity to deal with the scale of the predicted changes may be lacking:

> Bangladesh Climate Change Strategy and Action Plan also included the climate change issues, not only from the vulnerability perspective, but also from the capability perspective. Now we are talking about resilience.
>
> (Workshop participant, Dhaka)

Bangladesh Climate Change Resilience Fund

To assist Bangladesh with its adaptation strategies, the Bangladesh Climate Change Resilience Fund (BCCRF) was established in 2010 by the government of Bangladesh, development partners and the World Bank. Its objective is to build resilience amongst Bangladeshis and to support the implementation of the BCCSAP. To date it has distributed approximately US $170 million in funding for adaptation projects.

NGO adaptation frameworks

Several NGOs have also developed adaptation strategies to assist Bangladesh. These tend to be localized and based on community action research.

Box A.2 Community-level strategies adopted by NGOs in Bangladesh

- environmental management;
- water resource management;
- infrastructure development;
- livelihood diversification;
- social protection;
- education;
- agricultural research and development;
- fisheries research and development;
- governance structures such as social services; and
- disaster risk management including gender-sensitive strategies such as
- providing separate rooms for women in cyclone shelters; mini-shelters closer to villages;
- mobile medical teams in char areas; and using female voices in early warning announcements

(World Bank 2010: 61)

Despite the widespread acceptance of gender sensitivity in these actions, there is concern that government and NGOs are uncritically accepting inequitable gender roles and responsibilities and that gender inequalities are becoming further entrenched through actions, funding and outreach services (Neelormi 2010).

Of significance is that, despite global interactions, the plans and actions currently underway, and the support of INGOs, those at village level note that addressing basic needs is the most critical issue for survival. Local people note that adaptation efforts should focus on ensuring safe drinking water, sanitary latrines, a safe home and a stable income:

> No one can start worrying about adapting for the future while their immediate basic needs are not being met.
>
> (Raihan *et al.* 2010: 54)

Bibliography

Absar, S.S. (2003) 'Health hazards and labour laws in Bangladesh: A narrative-based study on women garment workers', *Asian Journal of Social Science*, 31(3): 452–477.

Acid Survivors Foundation (2011) *Statistics* (online). Available from: www.acid survivors.org/statistics.html [20 July 2011].

Adger, W.N., Agrawala, S., Mirza, M.M.Q., Conde, C., O'Brien, K., Pulhin, J., Pulwarty, R., Smit, B. and Takahashi, K. (2007) 'Assessment of adaptation practices, options, constraints and capacity', *Climate Change 2007: Impacts, Adaptation and Vulnerability. Contribution of Working Group II to the Fourth Assessment Report of the Intergovernmental Panel on Climate Change*, M.L. Parry, O.F. Canziani, J.P. Palutikof, P.J. van der Linden and C.E. Hanson (eds), Cambridge: Cambridge University Press, pp. 717–743.

Advancing Public Interest Trust (APIT) (2009) *A Study on Change in Women's Life after Sidr – Based on a Case Study – OGB after Sidr 2007 in Bangladesh*, Oxford: Oxfam.

Agarwal, B. (2010) *Gender and Green Governance: The political economy of women's presence within and beyond community forestry*, Oxford: Oxford University Press.

Agrawal, A., Perrin, N., Chatre, A., Benson, C. and Kononen, M. (2009) 'Climate Policy Processes, Local Institutions and Adaptation Actions: Mechanisms of translation and influence,' *Social Development Papers: Social Dimensions of Climate Change*, Paper No.119, Washington, DC: World Bank.

Ahmed, A.U. (2008) *Assessment of Vulnerability to Climate Change and Adaptation Options for the Coastal People of Bangladesh*, Dhaka: Practical Action.

Ahmed, A.U. (2006). Bangladesh Climate Change Impacts and Vulnerability: A Synthesis, *Climate Change Cell*, Dhaka: Bangladesh Department of Environment.

Ahmed, A.U., Neelormi, S. and Adri, N. (2007) *Entrapped in a Water World: Impacts of and adaptation to climate change induced water logging for women in Bangladesh*, Dhaka: Centre for Global Change.

Ahmed, A.U., Neerlormi, S., Adri, N., Alam, S. and Nuruzzaman, K. (2008) 'Climate Change, Gender and Vulnerable Groups in Bangladesh', *Climate Change Cell*, DoE, MoEF; Component 4b, CDMP, MoFDM, Dhaka.

Ahmed, F.E. (2004) 'The rise of the Bangladesh garment industry: Globalization, women workers, and voice', *NWSA Journal*, 16(2): 34–45.

Ahmed, M.R. (1992) 'Unseen workers: A sociocultural profile of women in Bangladesh agriculture', *Society & Natural Resources*, 5(4): 375–390.

Ahmed, S.M., Adams, A., Chowdhury, M. and Bhuiya, A. (2000) 'Gender, socio-economic development and health-seeking behaviour in Bangladesh', *Social Science and Medicine*, 51: 361–371.

Ahmed, S.N. (2005) 'Intimate partner violence against women: Experiences from a woman-focused development programme in Matlab, Bangladesh', *Journal of Health and Population Nutrition*, 23(1): 95–101.

Ahmed, S.S. and Bould, S. (2004). '"One able daughter is worth 10 illiterate sons": Reframing the patriarchal family', *Journal of Marriage and Family*, 66(5): 1332–1341.

Ahmed, S.Y., Chowdhury, M. and Bhuiya, A. (2001) 'Micro-credit and emotional well-being: Experience of poor rural women from Matlab, Bangladesh', *World Development*, 29(11): 1957–1966.

Aklimunnessa, K., Khan, M.M.H., Kabir, M. and Mori, M. (2007) 'Prevalence and correlates of domestic violence by husbands against wives in Bangladesh: Evidence from a national survey', *Journal of Men's Health and Gender*, 4(1): 52–63.

Alam, K., Shamsuddoha, MD., Tanner, T., Sultana, M., Huq, M.J. and Kabir, S.S. (2011) 'The political economy of climate resilient development planning in Bangladesh', *IDS Bulletin*, 42(3): 52–61.

Alam, M. (2004) 'Adverse Impacts of Climate Change on Development of Bangladesh: Integrating Adaptation into Politics and Activities,' *CLACC Working Paper No.1*. Dhaka: Bangladesh Centre for Advanced Studies.

Alam, S. (1985) 'Women and poverty in Bangladesh', *Women's Studies International Forum*, 8(4): 361–371.

Alam, M. and Golam Rabbani, M.D. (2007) 'Vulnerabilities and responses to climate change for Dhaka', *Environment and Urbanisation*, 19(1): 81–97.

Alston, M. (2013a) 'Women and adaptation', *WIREs Clim Change*. doi: 10.1002/wcc.232.

Alston, M. (2013b) 'Gender mainstreaming and climate change', *Women's Studies International Forum*, 21 March doi: 10.1016/j.wsif.

Alston, M. (2013c) 'Environmental social work: Accounting for gender in climate disasters', *Australian Social Work*, 66(2): 218–233.

Alston, M. (2013d) 'Gender mainstreaming', *A report for UNEP*, Geneva: PCDMB.

Alston, M. (2012) 'Gender and climate change in Australia and the Pacific', *Research, Action and Policy: Addressing the gendered impacts of climate change*, Margaret Alston and Kerri Whittenbury (eds), New York: Springer, pp. 175–188.

Alston, M. (2010) 'Gender and climate change in the Pacific', a report prepared for UNESCO, Monash University, Department of Social Work, Melbourne.

Alston, M. and Whittenbury, K. (2012) 'Does climatic crisis in Australia's food bowl create a basis for change in agricultural gender relations?' *Agriculture and Human Values*.

Alston, M. and Vize, S. (2010). *Gender and Climate Change in the Pacific*, Gender, Leadership and Social Sustainability (GLASS) research unit, Melbourne: Monash University.

Alston, M. and Mason, R. (2008). 'Who determines access to Australia's water? Gender, citizenship and stakeholder priorities in the Australian water crisis',*Rural Society* – Special edition, Gender and Water, 18(3): 214–219.

Alston, M., Whittenbury, K. and Haynes, A. (2011) 'The social impacts of declining water availability and ongoing drought in the Murray-Darling Basin: Short report', Melbourne: Monash University Department of Social Work.

Ambrus, A., Field, E. and Torero, M. (2010) 'Muslim family law, prenuptial agreements, and the emergence of dowry in Bangladesh', *The Quarterly Journal of Economics*, August: 1349–1397.

Amin, S. (1998) 'Family structure and change in rural Bangladesh', *Population Studies*, 52(2): 201–213.

Amin, S. (1997) 'The poverty-purdah trap in rural Bangladesh: Implications for women's roles in the family', *Development and Change*, 28: 213–233.

Aminuzzaman, Salahuddin M. (2010) *Local Government and Development in Bangladesh: Lessons learned and challenges for improving service delivery of Union Parishad*, Government of the Peoples' Republic of Bangladesh: Ministry of Local Government, Rural Development and Cooperatives.

Anwar, I., Killewo, J., Mahbub-E-Elahi, Chowdhury, K. and Dasgupta, S.K. (2004) 'Bangladesh: Inequalities in utilisation of maternal health care services – evidence from Matlab', *Health, Nutrition and Population Discussion Paper, Reaching the Poor Program Paper No.2*, Washington, DC: The World Bank.

Arends-Keunning, M. and Amin, S. (2001) 'Women's capabilities and the right to education in Bangladesh', *International Journal of Politics, Culture, and Society*, 15 (1): Risks and Rights in the 21st Century: Papers from the Women and Gender in Global Perspectives Program Symposium, University of Illinois at Urbana-Champaign, October 2000, pp. 125–142.

Arunachalam, R. and Naidu, S. (2006) 'Do Men Pay for Fertility? Marriage market effects of a family planning experiment in Bangladesh', Working Paper, Berkeley: University of California.

Asian Human Rights Commission (2011) *The State of Human Rights. Human rights report in Bangladesh for the year 2010* (online). Available from: www.humanrights.asia/resources/hrreport/2010/1%20Bangladesh_2010.pdf [25 July 2011].

Asian Indigenous People's Pact (AIPP) (2007) *A Brief Account of Human Rights Situation of the Indigenous Peoples in Bangladesh*, Chiang Mai: Asian Indigenous People's Pact.

Asian Women's Resource Exchange (AWORC) (1998) 'CEDAW and women: The Bangaldesh Experience', available from: www.aworc.org/bpfa/pub/sec_i/hum00001.html [18 August 2013].

Associated Press (AP) (2013) 'Bangladeshi mourners gather at wreckage of garment factory', available from: www.guardian.co.uk/world/2013/may/14/bangladesh-mourners-garment-factory [29 July 2013].

Azim, S. (2001) *Naripokkho's Pilot Study on Violence against Women in Bangladesh*, Dhaka, Naripokkho.

Bachelet, M: (2012) *Keynote Remarks of Michelle Bachelet at Rio+20 Corporate Sustainability Forum, Gender Equality for Sustainability session*, available from: www.unwomen.org/2012/06/keynote-remarks-of-michelle-bachelet-at-rio20-corporate-sustainability-forum-gender-equality-for-sustainability-session.

Bangla 2000 (2013) *Bangladesh portal*, available from: www.bangla2000.com/index.asp [4 August 2013].

Bangladesh Bureau of Statistics (BBS) (2011). *Census Reports* (online). Available from: www.bbs.gov.bd [1 June 2011].

Bangladesh Bureau of Statistics (BBS) (2010) *Area, Yield Rates and Production of Major Crops 2007–10*, available from: www.bbs.gov.bd/WebTestApplication/userfiles/Image/AgricultureCensus/All_Crops_summary_09-10.pdf [4 August 2013].

Bangladesh Bureau of Statistics (BBS) (2009) *Facts and Figures of Gender Compendium of Bangladesh 2009 (In Shortcut)* (online). Available from: www.bbs.gov.bd/WebTestApplication/userfiles/Image/SubjectMatterDataIndex/GSCompend_09.pdf [26 May 2011].

Bangladesh Bureau of Statistics (BBS) (2008) *Statistical Year Book of Bangladesh 2006*, Dhaka: BBS.

Bangladesh Garment Manufacturers and Exports Association (BGMEA) (2009) *BGMEA at a Glance* (online). Available from: www.bgmea.com.bd/home/pages/aboutus [1 June 2011].

Bangladesh Institute of Development Studies (BIDS) (2004) *Baseline Survey for Assessing Attitudes and Practices of Male and Female Members and In-laws towards Gender-based Violence*, Dhaka: United Nations Population Fund (UNFPA).

Bangladesh News (2008) *BNMS Calls for Immediate Action on Women's Policy*, 25 April, available from: www.independent-bangladesh.com/200804244742/country/bnms-calls-for-immediate-govt-action-on-womens-policy.html [23 August 2013].

Bartky, S. L. (1990) *Feminity and Domination: Studies in the phenomonology of oppression*, New York: Routledge.

Bates, L.M., Maselko, J. and Schuler, S.R. (2007) 'Women's education and the timing of marriage and childbearing in the next generation: Evidence from rural Bangladesh', *Studies in Family Planning*, 38(2): 101–112.

Bates, L.M., Schuler, S.R., Islam, F. and Islam, K. (2004). 'Socioeconomic factors and processes associated with domestic violence in rural Bangladesh', *International Family Planning Perspectives*, 30(4): 190–199.

Bauman, Z. (2011) *Collateral Damage: Social inequalities in a global age*, Malden, MA: Polity.

Beck, T. (2005) 'Learning Lessons from Disaster Recovery: The case of Bangladesh,' Disaster Risk Management Working Paper Series, No.11, Washington, DC: World Bank.

Beneria, L. (2003) *Gender, Development and Globalisation: Economics as if people mattered*, New York: Routledge.

Benson, C. and Clay, E.J. (2004) 'Understanding the economic and financial impacts of natural disasters', *Disaster Risk Management Series, No.4*, Washington, DC: The World Bank.

Berik, G. and van der Muelen Rodgers, Y. (2008) *Options for Enforcing Labor Standards: Lessons from Bangladesh and Cambodia. Working Paper Series, No.14*, Salt Lake City: University of Utah Department of Economics.

Bernasek, A. (2003) 'Banking on social change: Grameen Bank lending to women', *International Journal of Politics, Culture, and Society*, 16(3): 369–385.

Bhuiya, A., Sharmin, T. and Hanifi, S.M.A. (2003) 'Nature of domestic violence against women in a rural area of Bangladesh: Implication for preventive interventions', *Journal of Health Population and Nutrition*, 21(1): 48–54.

Bhuiya, A.A., Chowdhury, M.R., Momen, M. and Khatan, M. (2005) 'Marital disruption: Determinants and consequences on the lives of women in a rural area of Bangladesh', *Journal of Health, Population and Nutrition*, 23(1): 82–94.

Bhuyan, A. R. (2005) 'Millennium Development Goals (MDGs): A review of Bangladesh's achievements', available from: http://ibtra.com/pdf/journal/v2_n1_article4.pdf [23 August 2013].

Blanchett, T. (2001) *Lost Innocence, Stolen Childhood*, Dhaka: University Press.

Blunch, N-H. and Das, M.B. (2007) 'Changing Norms about Gender Equality in Education: Evidence from Bangladesh,'. Policy Research Working Papers No.4404, Washington, DC: World Bank.
Bord, R.J. and O'Connor, R.E. (1997) 'The gender gap in environmental attitudes: The case of perceived vulnerability to risk: Research on the environment', *Social Science Quarterly* 78(4): 830–840.
Bordo, S. (1989) 'The body and the reproduction of feminity, *Gender/Body/Knowledge: Feminist reconstructions of being and knowing*, Alison Jaggar and Susan Bordo (eds), New Brunswick, NJ: Rutgers University Press.
Bose, M.L., Ahmad, A. and Hossain, M. (2009) 'The role of gender in economic activities with special reference to women's participation and empowerment in rural Bangladesh', *Gender Technology and Development*, 13(1): 69–102.
Bourdieu, P. (2001) *Masculine Domination*, Cambridge: Polity Press.
Bourdieu, P. (1992) *Language and Symbolic Power*, Cambridge: Polity Press.
Bourdieu, P. (1970) with Jean-Claude Passeron (in French: *La Reproduction. Éléments pour une théorie du système d'enseignement*, Les Éditions de Minuit) *Reproduction in Education, Society and Culture* (Theory, Culture and Society Series), Thousand Oaks, CA: Sage.
Brandt, M. and Kaplan, J.A. (1995) 'The tension between women's rights and religious rights: Reservations to Cedaw by Egypt, Bangladesh and Tunisia', *Journal of Law and Religion*, 12(1): 105–142.
BRIDGE (2008) *Gender and Climate Change: Mapping the linkages. A scoping study on knowledge and gaps*, Brighton: BRIDGE.
Brouwer, R., Akter, S., Robander, L. and Haque, E. (2007) 'Socioeconomic vulnerability and adaptation to environmental risk: A case study of climate change and flooding in Bangladesh', *Risk Analysis*, 27(2): 313–326.
Cain, M. (1978) 'The household life cycle and economic mobility in rural Bangladesh', *Population and Development Review*, 4(3): 421–438.
Cain, M., Khanam, S.R. and Nahar, S. (1979) 'Class, patriarchy and women's work in Bangladesh', *Population and Development Review*, 5(3): 405–438.
Cannon, T. (2002) 'Gender and climate hazards in Bangladesh', *Gender and Development*, 10(2): 45–50.
Centre for Environment and Geographic Information Services (2008) Quoted by Mark Dummett in *BBC News: Bangaldesh Land Mass 'Is Growing'*, 30 July, available from: http://news.bbc.co.uk/2/hi/south_asia/7532949.stm [16 July 2013].
Centre for Policy Dialogue (CPD) (2008a) *Bangladesh's Apparel Sector in Post-MFA Period: A benchmarking study on the ongoing restructuring process*, Dhaka: Centre for Policy Dialogue.
Centre for Policy Dialogue (CPD) (2008b) *Gender and Trade Liberalization in Bangladesh: The case of the ready-made garments*, Dhaka: Centre for Policy Dialogue.
Centre for Policy Dialogue (CPD) (2009) *Report No.98: Achieving Universal Primary Education with Quality and Equity*, Dhaka: Centre for Policy Dialogue.
Chemin, M. (2008) 'The benefits and costs of microfinance: Evidence from Bangladesh', *Journal of Development Studies*, 44(4): 463–484.
Chittagong Hill Tracts Development Facility (CHTDF) (2011). *Home* (online). Available from: www.chtdf.org/index.php/cht-issues/cht-situation [26 July 2011].
Chittagong Hill Tracts Development Facility (CHTDF) (2009) *Socio-economic Baseline Study of Chittagong Hill Tracts*, Dhaka: United Nations Development Programme.

Chittagong Hills Tracts Commission (CHTDF) (1991) *'Life Is Not Ours': Land and human rights in the Chittagong Hill Tracts, Bangladesh*, Copenhagen: IWGIA.

Chowdhury, F. (2010) 'Dowry, women and law in Bangladesh', *International Journal of Law, Policy and the Family*, 24(2): 198–221.

Chowdhury, F. (2009) 'Theorising patriarchy: The Bangladesh context', *Asia Journal of Social Science*, 37: 599–622.

Chowdury, K. R. (2013): *In Bangladesh a Stand for Women's Rights*, available from: http://khabarsouthasia.com/en_GB/articles/apwi/articles/features/2013/04/23/feature-01 [18 August 2013].

Chowdhury, M.E., Ronsmans, C., Killewo, J., Anwar, I., Gausia, K., Das-Gupta, S., Blum, L.S., Dieltiens, G., Marshall, T., Saha, S. and Borghi, J. (2006) 'Equity in use of home-based or facility-based skilled obstetric care in rural Bangaldesh: An observational study', *The Lancet*, 367: 327–332.

Christensen, J.H., Hewitson, B., Busuioc, A., Chen, A., Gao, X., Held, I., Jones, R., Kolli, R.K., Kwon, W.-T., Laprise, R., Magaña Rueda,V., Mearns, L., Menéndez, C.G., Räisänen, J., Rinke, A., Sarr, A. and Whetton, P. (2007) 'Regional climate projections', *Climate Change 2007: The physical science basis. Contribution of Working Group I to the Fourth Assessment Report of the Intergovernmental Panel on Climate Change*, S. Solomon, D. Qin, M. Manning, Z. Chen, M. Marquis, K.B. Averyt, M. Tignor and H.L. Miller (eds), Cambridge and New York: Cambridge University Press.

Chughtai, S. (2008) 'One year after Cyclone *Sidr*: Fear replaces hope', *Oxfam Briefing Notes*, November, Oxford: Oxfam International.

CIA World Fact Book (2013) *Bangladesh Demographic Profile*, available from: www.indexmundi.com/bangladesh/demographics_profile.html [25 July 2013].

Connell, R.W. (2011) *Confronting Equality: Gender, knowledge and global change*, Sydney: Allen & Unwin.

Connell, R.W. (2002) *Gender, Polity Press*, Cambridge and Malden, MA: Blackwell Publishers.

Connell, R.W. and Messerschmidt, J.W. (2005) 'Hegemonic masculinity: Rethinking the concept', *Gender & Society*, December, 19(6): 829–859.

Corral, T. (2010) 'Gender perspectives in adaptation strategies: The case of Pintadas Solar in the semi-arid region of Brazil, *Gender and Climate Change: An introduction*, Irene Dankelman (ed.), London: Earthscan pp. 138–144.

Dalal, K., Rahman, F. and Jansson, B. (2009) 'Wife abuse in rural Bangladesh', *J Biosoc Sci*, 41: 561–573.

Dankelman, I. (2010) 'Introduction: Exploring gender, environment and climate change',*Gender and Climate Change: An introduction*, Irene Dankelman (ed.), London: Earthscan, pp. 1–20.

Dankelman, I. and Jansen, W. (2010) 'Gender, environment and climate change: Understanding the linkages, *Gender and Climate Change: An introduction*, Irene Dankelman (ed.), London and Sterling, VA: Earthscan, pp. 21–54.

Dasgupta, S., Huq, M., Khan, Z.H., Ahmed, M.M.Z., Mukherjee, N., Khan, M.F. and Pandey, K. (2010) 'Vulnerability of Bangladesh to cyclones in a changing climate: Potential damages and adaptation cost', *Policy Research Working Paper, Paper 5280*, Washington, DC: World Bank.

De Haan, F., Allen, M., Purvis, J. and Daskalova, K. (2013) 'Introduction', *Women's Activism: Global perspectives from the 1890s to the present*, Francisca De Haan, Allen, Margaret, Purvis, June and Daskalova, Krassimira (eds), London: Routledge.

Debnath, A. (2009) 'The Bangladesh genocide: The plight of women', *Plight and fate of women during and following genocide*, S. Totten (ed.), New Brunswick, NJ: Transaction pp. 47–66.

Decker, M.R., Miller, E., Kapur, N.A., Gupta, J., Raj, A. and Silverman, J.G. (2008) 'Intimate partner violence and sexually transmitted disease symptoms in a national sample of married Bangladeshi women', *International Journal of Gynecology and Obstetrics*, 100: 18–23.

del Ninno, C. and Borsch, P.A. (2003) 'Public policy, markets and household coping strategies in Bangladesh: Avoiding a food security crisis following the 1998 floods', *World Development*, 31(7): 1221–1238.

Denton, F. (2002) 'Climate change vulnerability, impacts and adaptation: Why does gender matter?', *Gender and Development*, 10(2) July: 10–20.

Develtere, P. and Huybrechts, A. (2005) 'The impact of microcredit on the poor in Bangladesh', *Alternatives*, 30: 165–189.

Disaster Management Information Centre (2007) *Cyclone Situation Report Sidr*, Dhaka: Department of Disaster Management.

Do, Q-T., Iyer, S. and Joshi, S. (2006) *The Economics of Consanguineous Marriages. World Bank Policy Research Working Paper Series, No.4085* [online]. Available from: http://econ.worldbank.org [25 May 2011].

Dowding, K. (2006) 'Three-dimensional power: A discussion of Steven Lukes' power: A radical view,' *Political Studies Review*, 4: 136–145.

Economic and Social Commission for Asia and the Pacific (ESCAP) and United Nations Development Programme (UNDP) (2012) *Accelerating Equitable Achievement of the MDGs: Closing gaps in health and nutrition outcomes. Asia-Pacific Regional MDG Report*. Available from: www.unescap.org/pdd/calendar/CSN-MDG-NewDelhi-Nov-2011/MDG-Report2011-12.pdf [23 August 2013].

Enarson, E. (2012) *Women Confronting Natural Disaster: From vulnerability to resilience*, Boulder, CO: Lynne Reinner Publishers.

Enarson, E. (2009) *Women, Gender and Disaster*, Thousand Oaks, CA: Sage Publications.

Enarson, E. (2006) 'Women and lirls Last?: Averting the second post-*Katrina* disaster.' Available from: http://understandingkatrina.ssrc.org/Enarson [14 March 2011].

Feldman, S. and McCarthy, F. (1983) 'Purdah and changing patterns of social control among rural women in Bangladesh', *Journal of Marriage and Family*, 45(4): 949–959.

Ferree, M. (2012) *Varieties of Feminism: German gender politics in global perspective*, Stanford, CA: Stanford University Press.

Ferree, M. (2006) 'Globalisation and feminism: Opportunities and obstacles for activism in the global arena, *Global Feminism: Transnational women's activism, organising and human rights*, Myra Marx Ferree and Aili Mari Tripp (eds), New York: New York University Press, pp. 3–23.

Ferree, M. M. and Tripp, A. M. (2006) *Global Feminism: Transnational women's activism, organising and human rights*, New York: New York University Press.

Food and Agricultural Organization (FAO) (2007a) *Gender and Climate Change: Existing research and knowledge gaps*, Rome: Gender and Population Division, FAO.

Food and Agricultural Organization (FAO) (2007b) *Women, Agriculture and Food Security*, Rome, FAO, available from: www.fao.org/worldfoodsummit/english/fsheets/women.pdf [22 December 2012].

Food and Agricultural Organization (FAO) (nd) *Gender Equality*, available from: ftp://ftp.fao.org/docrep/fao/011/i0765e/i0765e10.pdf [16 January 2013].

Foucault, M. (1991) 'Governmentality', *The Foucault Effect: Studies in governmentality*, Graham Burchell, Colin Gordon and Peter Miller (eds), Chicago, IL: University of Chicago Press, pp. 85–103.

Foucault, M. (1998). *The History of Sexuality: The will to knowledge*, London: Penguin.

Franzway, S., Court, D. and Connell, R.W. (1989) *Staking a Claim: Feminism, bureaucracy and the state*, Cambridge: Polity Press.

Garcia-Moreno, C., Jansen, A.F.M., Ellsberg, M., Heise, L. and Watts, C. (2005) *WHO Multi-country Study on Women's Health and Domestic Violence against Women, World Health Organization: Initial results on prevalence, health outcomes and women's responses*, Geneva: World Health Organization Press.

Gierbo, H.C. and Imam, N. (2006) *The Motivations behind Giving and Taking Dowry. BRAC Research Monograph Series, No.28*, Dhaka: BRAC Research and Evaluation Division.

Giddens, A. (2009) *The Politics of Climate Change*, Cambridge: Polity Press.

Global Poverty Project (2013) *Global Poverty Info Bank*, available from: www.globalpovertyproject.com/infobank/women [15 January 2013].

Goldsmith, R., Feygina I., and Jost, J. (2012) 'The gender gap in environmental attitudes, *Research, Action and Policy: Addressing the gendered impacts of climate change*, M. Alston, and K. Whittenbury (eds), New York: Springer, pp. 159–171.

Government of Bangladesh (2009). *Bangladesh Climate Change Strategy and Action Plan 2009*, Dhaka: Ministry of Environment and Forests.

Government of Bangladesh (GoB) (2005) *National Adaptation Programme of Action (NAPA)*, Dhaka: Ministry of Environment and Forest Government of the People's Republic of Bangladesh.

Green Left Weekly (2006) 'Bangladesh Women Organise for their Rights', available from: www.wluml.org/node/2982 [20 August 2013].

Guhathakurta, M. and van Schendel, W. (eds) (2013) *The Bangladesh Reader History, Culture, Politics*, The World Series, Durham, NC: Duke University Press.

Hadi, S.T. (2009) '"It's OK to beat my wife?" – Patriarchal perceptions of Bangladeshi respondents and factors associated', *Bangladesh e-Journal of Sociology*, 6(2): 4–14.

Hadi, S.T. (2000) 'Prevalence and correlates of the risk of marital sexual violence in Bangladesh', *Journal of Interpersonal Violence*, 15: 787–805.

Hagerty, D. T. (2008) 'Bangladesh in 2007: Democracy interrupted, political and environmental challenges ahead', *Asian Survey*, 48(1): January/February 177–183.

Haraway, D. (1990) 'A manifesto for cyborgs: Science, technology and socialist feminism in the1980s', reprinted in Linda J. Nicholson (ed.), *Feminism/Postmodernism*, London: Routledge.

Haraway, D. (1988) 'Situated knowledges: The science question in feminism and the privilege of partial perspective,' *Feminist Studies*, 14(3): 575–599.

Harding, S. (1992) *Whose Science? Whose Knowledge? Thinking From Women's Lives*, Ithaca, NY: Cornell University Press.

Harris, R. (2010) 'Gender aspects of climate change in the US Gulf Coast Region', *Gender and Climate Change: An introduction*, Irene Dankelman (ed.), London: Earthscan, pp. 152–157.

Hashemi, S.M., Schuler, S.R. and Riley, A.P. (1996) 'Rural credit programs and women's empowerment in Bangladesh', *World Development*, 24(4): 635–653.

Hausmann, R., Tyson, L.D. and Zahidi S. (2007) *The Global Gap Report*, Geneva: World Economic Forum.

Hawthorne, S. (2001) *Wild Politics*, North Melbourne: Spinifex Press.

Highbeam Research (2005)., *Bangladesh*, available from: www.infoplease.com/country/bangladesh.html [22 July 2013].

Hindin, M.J., Kishor, S. and Ansara, D.L. (2008) *Intimate Partner Violence amongst Couples in 10 DHS Countries: Predictors and health outcomes*, Measure DHS Analytical Studies No.18, Calverton: USAID.

Hoque, M. and Itohara, Y. (2008). 'Participation and decision making role of rural women in economic activities: A comparative study for members and non-members of the micro-credit organisations in Bangladesh', *Journal of Social Sciences*, 4(3): 229–236.

Hossain, H. (2006) 'Women's movements in Bangladesh: The struggle within,' 5 January, available from: www.europe-solidaire.org/spip.php?article1567 [20 August 2013].

Hossain, K. (2009) 'Birth of a climate change policy and related debates: Analysing the case of Bangladesh'. Paper presented at Environmental Policy: A multinational conference on policy analysis and teaching methods, 11–13 June, Seoul, South Korea.

Hossain, K. (2003) 'In search of equality: Marriage related laws for Muslim women in Bangladesh', *Journal of International Women's Studies*, 5(1): 96–113.

Hossain, K.T. (2007) 'Violence against women in Bangladesh.' Proceeding of the 8th Annual Conference of Hong Kong Sociological Association Narratives and Perspectives in Sociology: Understanding the past, envisaging the future, C.S. Keung, H. Traverand L. Xiuguo (eds), pp. 257–264.

Hossain, M.B., Phillips, J.F. and Pence, B. (2006) 'The effect of women's status on infant and child mortality in four rural areas of Bangladesh', *J Biosoc Sci*, 39: 355–366.

Hossain, N. (2004) *Access to Education for the Poor and Girls: Educational achievements in Bangladesh*, Washington, DC: The World Bank.

Huda, S. (2006) 'Dowry in Bangladesh: Compromising women's rights', *South Asia Research*, 26: 249–268.

Human Rights Watch (2011) *World Report 2011: Events of 2010*, New York: Human Rights Watch.

Hunt, J. and Kasynathan, N. (2002) 'Reflections on microfinance and women's empowerment', *Development Bulletin*, 57: 71–75.

Hurt, L.S., Ronsmans, C. and Saha, S. (2004) 'Effects of education and other socio-economic factors on middle age mortality in rural Bangladesh', *Journal of Epidemiology and Community Health*, 58: 315–320.

Independent Commission for Aid Impact (ICAI) (2011) *The Department for International Development's Climate Change Impact in Bangladesh. Report 3*, November. Available from: www.oecd.org/countries/bangladesh/49092047.pdf [23 August 2012].

Intergovernmental Panel on Climate Change (IPCC) (2014) *Fifth Assessment Report: Climate change, impacts, adaptation and vulnerability*, available from: www.ipcc.ch/report/ar5/wg2, [10 April 2014].

Intergovernmental Panel on Climate Change (IPCC) (2007). 'Fourth Assessment Report: Climate change 2007,' Working Group II: Impacts, adaption and vulnerability, Geneca: IPCC.

Intergovernmental Panel on Climate Change (IPCC) (1991) *Climate Change: First Assessment Report*, Cambridge: Cambridge University Press.

International Fund for Agricultural Development (IFAD) (2007) *Gender and Water*, Rome: IFAD.

International Women's Democracy Centre (2008) *Fact Sheet, Women's Political Participation*, available from: www.iwdc.org/resources/fact_sheet.htm [15 January 2013].

Islam, K. S. (2012) 'Breaking down the Birangona: Examining the (divided) media discourse on the war heroines of Bangladesh's independence movement', *International Journal of Communications*, 30(August): 21–31.

Islam, M.M. (1999). 'Adolescent childbearing in Bangladesh', *Asia-Pacific Population Journal*, 14(3): 73–87.

Jahan, R. and Amundsen, I. (2012) *The Parliament of Bangladesh: Representation and accountability.* CPD-CMI Working Paper 2. Centre for Policy Dialogue, Dhaka, www.cpd.org.bd/blog and Centre for Policy Dialogue and Chr Michelsen Institute, Norway.

Jahanbegloo, R. and Vandana S. (2012) *Talking Environment: Vandana Shiva in conversation with Ramin Jahanbegloo.* Oxford: Oxford University Press.

Jayaraman, A., Mishra, V. and Arnold, F. (2008) 'The effect of family size and composition on fertility desires, contraception adoption, and method choice in South Asia', *DHS Working Papers Series, No.40*, Calverton: Macro International.

Jennings, S. and McGrath, J. (2009) *What Happened to the Seasons?*, Oxford: Oxfam Great Britain.

Johnson, K.B. and Das, M.B. (2009) 'Spousal violence in Bangladesh as reported by men: Prevelance and risk factors', *Journal of Interpersonal Violence*, 24(6): 977–995.

Kabeer, N. (2011) 'Between affiliation and autonomy: Navigating pathways of women's empowerment and gender justice in rural Bangladesh', *Development and Change*, 42(2): 499–528.

Kabeer, N. (2001) 'Conflicts over credits: Re-evaluating the empowerment potential of loans to women in rural Bangladesh', *World Development*, 29: 63–84.

Kabeer, N. (1997) 'Women, wages and intra-household power relations in urban Bangladesh', *Development and Change*, 28: 361–302.

Kabeer, N. and Mahmud, S. (2004) 'Globalisation, gender and poverty: Bangladeshi women workers in export and local markets', *Journal of International Development*, 16: 93–109.

Kabir, A., Jahan, G. and Jahan, R. (2001). 'Female age at marriage as a determinant of fertility', *The Sciences*, 1(6): 372–376.

Kabria, N. (1995) 'Culture, social class, and income control in the lives of women garment workers in Bangladesh', *Gender and Society*, 9(3): 289–309.

Kamrunnahar, M. (2007) 'Domestic violence against women in Bangladesh: A study of Rajshahi City', *Yokohama Journal of Technology Management Studies*, 6: 51–61.

Kelly, P. M. and W. N. Adger (2009) 'Theory and practice in assessing vulnerability to climate change and facilitating adaptation,' *Adaptation to Climate Change*, E. Lisa F. Schipper and Ian Burton (eds), London: Earthscan pp. 161–186.

Khan, I.K. (2000) *Socio-economic Profile of Sex Workers in Chittagong. CPD-UNFPA Publication Series, Paper 11*, Dhaka: Centre for Policy Dialogue.

Khan, M.E., Rob, U. and Hossain, S.M. (2000). 'Violence against women and its impact on women's lives – Some observations from Bangladesh', *Journal of Family Welfare*, 46(2): 12–24.

Khan, S., Islam, A. and Haque, I. (1996) *Political Parties, Political Culture and the Deomocratic Transition in Bangladesh*, Bangladesh: University Press Limited.

Khandker, S.R. (2007) 'Coping with flood: Role of institutions in Bangladesh', *Agricultural Economics*, 36: 169–180.

Khandker, S.R. (2005) 'Micro finance and poverty: Evidence using panel data from Bangladesh', *The World Bank Economic Review*, 19(2): 263–286.

Khatun, F. and Islam, A.K.N.M. (2010) 'Policy Agenda for Addressing Climate Change in Bangladesh: Copenhagen and beyond.' Occasional Papers, Paper 88, Dhaka: Centre for Policy Dialogue.

Kibria, N. (1995) 'Culture, social class and income control in the lives of women garment workers in Bangladesh', *Gender and Society*, 9(3).

Ki-Moon, B. (2009). 'UN Secretary-General Press Release', available from: www.un.org/News/Press/docs/2009/sgsm12488.doc.htm [6 December 2011].

Kitkiri, K. (2011). Climate change and migration: A case study from rural Bangladesh', *Gender & Development*, 19(1): 23–38.

Koenig, M.A., Ahmed, S., Hossain, M.B. and Mozumder, A.B.M.K.A. (2003) 'Women's status and domestic violence in rural Bangladesh: Individual- and community-level effects', *Demography*, 40(2): 269–288.

Koenig, M.A., Jamil, K., Streatfield, P.K., Saha, T., Al-Sabir, A., Arifeen, S.E., Hill, K. and Haque, Y. (2007) 'Maternal health and help-seeking behaviour in Bangladesh: Findings from a national survey', *International Family Planning Perspectives*, 33(2): 75–82.

Kumar, P. and Narayan, S. (2007) 'Sink or Swim: Why disaster risk reduction is central to surviving floods in South Asia.' Oxfam International Briefing Note, 12 August, Oxford: Oxfam International.

Lambrou, Y. and Nelson, S. (2010) *Farmers in a Changing Climate: Food Security in Andhra Pradesh, India*, Rome: Food and Agricultural Organization of the United Nations (FAO).

Lane, R. and McNaught R. (2009) 'Building gendered approaches to adaptation in the Pacific,*Gender and Development*, 17(1): March 67–80.

Linden, M. (2013) 'Climate change will bring greater extremes in weather, warns Government's chief scientific adviser', *The Independent*, 25 March, available from: www.independent.co.uk/environment/climate-change/climate-change-will-bring-greater-extremes-in-weather-warns-governments-chief-scientific-adviser-8547893.html [30 April 2013].

Lindenbaum, S. (1981) 'Implications for women of changing marriage transactions in Bangladesh', *Studies in Family Planning*, 12(1): 394–401.

Lourdes, B. (2003) *Gender, Development, and Globalization: Economics as if people mattered*. New York: Routledge.

Lukes, S. (2005) *Power: A radical view*, 2nd edition, Basingstoke: Palgrave Macmillan.

McCright, A.M. (2010) 'The effects of gender on climate change knowledge and concern in the American public', *Population and Environment*, 32: 66–87.

McGranahan, G., Balk, D. and Anderson, B. (2006): 'Low coastal zone settlements', *Tiempo*, 59: 23–26.

McLaren, M. A. (2002) *Feminism, Foucault and Embodied Subjectivity*, Albany: State University of New York Press.

McMichael, A.J. (2003) 'Global climate change and health: An old story writ large', *Climate Change and Human Health: Risks and Responses*, A.J. McMichael, D.H. Campbell-Lendrum, C.F. Corvalan, C.F., K.L. Eloi, A. Githeko, J.D. Scheraga and A. Woodward, (eds), Geneva: World Health Organization.

McSmith, A. (2006) 'The rich nations pollute, and the poor nations suffer consequences', *The Independent*, March 25, London.

Maddison, S. and Franzway, S. (2006). 'Reconsidering staking a claim', *Australian Feminist Studies*, 21(50): 139–149.

Magrath, J. (2007) 'Climate Alarm: Disaster increases as climate change bites.' Oxfam Briefing Paper, No.108, Oxford: Oxfam International.

Mahmud, S. (2003) 'Actually how empowering is microcredit?', *Development & Change*, 34(4): 577–605.

Mannan, M.A. (2000) 'Female-headed Households in Rural Bangladesh: Strategies for well-being and survival.' CDP-UNFPA Publication Series, Paper 10, Dhaka: Centre for Policy Dialogue.

Masood, M.J. and Wahra, G.N. (1997) 'Bangladesh: Learning to deal with disasters', *South Asian Women: Facing disasters, securing life*, P. Fernando and V. Fernando (eds), Colombo: Duryog Nivaran Secretarial.

Mehta, M. (2004) 'Towards ending violence against women in South Asia', *Oxfam Briefing Papers*, No.66, Oxford: Oxfam International.

Mitchell, T., Tanner, T. and Lussier, K. (2007) *We Know What We Need: South Asian women speak out on climate change adaptation*, Johannesburg: Institute of Development Studies and ActionAid.

Mitra, S.N., Al-Sabir, A., Saha, T. and Kumar, S. (2004). *Bangladesh Demographic and Health Survey 2004*, Dhaka: NIPORT and Mitra Associates.

Mixmarket (2011) *Microfinance in Bangladesh* [online]. Available from: www.mixmarket.org/mfi/country/Bangladesh [15 July 2011].

Monsoor, T. (1999) *From Patriarchy to Gender Equity Family Law and Its Impact on Women in Bangladesh*, Dhaka: The University Press Limited.

Mookherjee, N. (2011) '"Never again": Aesthetics of "genocidal" cosmopolitanism and the Bangladesh Liberation War Museum,' *Journal of the Royal Anthropological Institute (N.S.)*, 17: S71–S91.

Moser, S.C. and Ekstrom, J.A. (2010) 'A framework to diagnose barriers to climate change adaptation', *PNAS*, 107(51): 22026–22031.

Mustafa, H. B. (2013) *Women's Forum Demand Shafi's Arrest*. Available from: http://news.priyo.com/2013/07/14/womens-forum-demand-shafis-arrest-80987.html [18 August 2013].

Nahar, Q. and Min, H. (2008). *Trends and Determinants of Adolescent Childbearing in Bangladesh*, DHS Working Paper Series, Calverton: USAID.

Nanda, P. (1999) 'Women's participation in rural credit programmes in Bangladesh and their demand for formal health care: Is there a positive impact?', *Health Economics*, 8(5): 415–428.

NAPA (2005) *National Adaptation Programme of Action (NAPA) Final Report*, Dhaka: Ministry of Environment and Forest, Government of the People's Republic of Bangladesh.

Nasreen, M. (2008) *Violence Against Women during Flood and Post-flood Situations in Bangladesh*, Dhaka: ActionAid Bangladesh.

National Institute of Population Research and Training (NIPORT), Mitra and Associates and Macro International (2009) *Bangladesh Demographic and Health Survey 2007*, Dhaka and Calverton, MD: National Institute of Population Research and Training, Mitra and Associates and Macro International, available from: www.measuredhs.com/pubs/pdf/FR207/FR207[April-10-2009].pdf [28 July 2013].

National Institute of Population Research and Training (NIPORT), Mitra and Associates and Macro International (2005). *Bangladesh Demographic and Health Survey 2004*. Dhaka, Bangladesh and Calverton, MD: National Institute of Population Research and Training, Mitra and Associates and Macro International.

National Institute of Population Research and Training (NIPORT), OCR Macro, Johns Hopkins University, ICDDR, B. (2003) *Bangladesh Maternal Health Services and Maternal Morbidity Survey 2001*, Dhaka and Calverton, MD.

Naved, R.T. and Persson, L.A. (2010). 'Dowry and spousal physical violence against women in Bangladesh', *Journal of Family Issues*, 31(6): 830–856.

Naved, R.T. and Persson, L.A. (2005) 'Factors associated with spousal physical violence against women in Bangladesh', *Studies in Family Planning*, 36(4): 289–300.

Naved, R.T., Newby, M. and Amin, S. (2001) 'The effects of migration and work on marriage of female garment workers in Bangladesh', *International Journal of Population Geography*, 7: 91–104.

Naved, R.T., Azim, S., Bhuiya, A. and Persson, L.A. (2006) 'Physical violence by husbands: Magnitude, disclosure and help-seeking behaviour of women in Bangladesh', *Social Science and Medicine*, 62: 2917–2929.

Neefjes, K. and Nelson, V. (2010) 'Responding to climate change in Vietnam: Opportunities for improving gender equality,' *Gender and Climate Change: An introduction*, Irene Dankelman (ed.), London: Earthscan, pp. 107–114.

Neelormi, S. (2010) 'Addressing gender issues in adaptation', *Reducing Vulnerability to Climate Change: The pioneering example of community-based adaptation*, A.U. Ahmed (ed.), Dhaka: Centre for Global Change and CARE Bangladesh, pp. 111–127.

Neumayer, E. and Pluemper, T. (2007) 'The gendered nature of natural disasters: The impact of catastrophic events on the gender gap in life expectancy 1981–2002', available from: http://ssm.com/abstract=874965.

Odigie-Emmanuel, O. (2010) 'The gender impact of climate change in Nigeria,' *Gender and Climate Change: An introduction*, Irene Dankelman (ed.), London: Earthscan, pp. 123–129.

Oxfam (2009) *The Future is Here: Climate change in the Pacific*, Victoria: Oxfam Australia, Carlton.

Oxfam GB and Naripad (2011) *Handbook: Women leadership in disaster risk management*, Dhaka: Oxfam GB.

Oxfam International (2008) *Rethinking Disasters*, New Delhi: South Asia Regional Centre, Oxfam (India) Trust.

Pakulski, J. (2004) *Globalizing Inequalities*, Sydney: Allen & Unwin.

Papreen, N., Sharma, A., Sabin, K., Begum, L., Abdulla, S.K.A. and Baqui, H. (2000): 'Living with infertility: Experiences among urban slum populations in Bangladesh', *Reproductive Health Matters*, 8(15): 33–44.

Paris, T.R., Chowdhury, A. and Bose, M.L. (2004) 'Changing Women's Roles in Homestead Management: Mainstreaming women in rural development.' CDP Occasional Papers Series, Paper 42, Dhaka: Centre for Policy Dialogue.

Parveen, S. and Leonhäuser, I.-U. (2004) 'Empowerment of rural women in Bangladesh: A household level analysis', Abstracts of German Tropentag, Humbold-University, Berlin, 5–7 October, p. 95 (online). Available from: www.tropentag.de/2004/abstracts/full/382.pdf [25 May 2011].

Pelling, M. (2011) *Adaptation to Climate Change: From resilience to transformation*, London: Routledge.

Pender, J.S. (2008) *What Is Climate Change? And How It Will Effect Bangladesh*? Briefing Paper (final draft), Church of Bangladesh Social Development Programme, Dhaka, available from: www.kirkensnodhjelp.no/Documents/Kirkens%20Nødhjelp/Publikasjoner/Temahefter/FINAL%20Draft%20WHAT%20IS%20CLIMATE%20

CHANGE%20AND%20HOW%20IT%20MAY%20AFFECT%20BANGLADESH.pdf [28 September 2013].

People's Republic of Bangladesh (2011) *The Constitution*, available from: http://bdlaws.minlaw.gov.bd/print_sections_all.php?id=367 [14 September 2014].

Pettengell, C. (2010) *Climate Change Adaptation: Enabling people living in poverty to adapt*, Oxford: Oxfam Great Britain.

Pitt, M.M. and Khandker, S.R. (1998). 'The impact of group-based credit programs in Bangladesh: Does the gender of participants matter?', *Journal of Political Economy*, 106(5): 958–996.

Pitt, M.M., Khandker, S.R. and Cartwright, J. (2003) 'Does Micro-credit Empower Women? Evidence from Bangladesh,' *Policy Research Working Paper* 2998, Washington, DC: The World Bank.

Plan International (2009) *Because I am a Girl: The state of the world's girls 2009. Girls in the global economy: Adding it all up*, Surrey: Plan International.

Rahman, A. (1994) 'Women, cultural ideology and change in rural Bangladesh', *Canadian Journal of Development Studies*, 15(3): 429–443.

Rahman, A.S. (1997) 'Gender equality, equity and empowerment of women in Bangladesh', *Population and Development Issues in Bangladesh*, A. Barkatand S.R. Howlader (eds), Dhaka: Ministry of Health and Family Welfare pp. 51–72.

Rahman, M. and Nasrin, S.O. (2009) 'Mothers' nutritional status in an impoverished nation: Evidence from rural Bangladesh', *The InternetJ Nutr and Wellness*, 7(1).

Rahman, S. and Routray, J.K. (1998) 'Technological change and women's participation in crop production in Bangladesh', *Gender Technology and Development*, 2: 243–267.

Rahman, S., Junankar, P.N. and Mallik, G. (2009) 'Factors influencing women's empowerment on microcredit borrowers; A case study in Bangladesh', *Journal of the Asia Pacific Economy*, 14(3): 287–303.

Raihan, M.S., Huq, M.J., Alsted, N.G. and Andreasen, M.H. (2010) *Understanding Climate Change from Below, Addressing Barriers from Above*, Dhaka: ActionAid Bangladesh.

Randerson, J. (2008) 'Food crisis will take hold before climate change, warns chief scientist', *The Guardian*, March 7, available from: www.guardian.co.uk/science/2008/mar/07/scienceofclimatechange.food [29 November 2011].

Rashid, M., Singha, D. and Imam, H. (2009) *Climate Change Vulnerability in Bangladesh: Strategic position of DSK/DCA in the field of climate change adaptation initiatives in Bangladesh*, Dhaka: Dustha Shastha Kendra.

Rashid, S.B. and Michaud, S. (2000) 'Female adolescents and their sexuality: Notions of honour, shame, purity and pollution during the floods', *Disasters*, 24(1): 54–70.

Renton, A. (2009) Suffering the Science: Climate change, people and poverty. Oxfam Briefing Paper, 6 July, Oxford: Oxfam International.

Riaz, A. (2004) 'Bangladesh in 2004: The politics of vengeance and the erosion of democracy, *Asian Survey*, 45(1) January/February: 112–118.

Riley, M. (2008) 'A feminist political economy framework. Centre of concern', available from: https://www.coc.org/files/Riley%20-%20FPE_0.pdf [4 July 2013].

Roach, R. (2005) *Dried Up, Drowned Out: Voices from the developing world on a changing climate*, Teddington: Tearfund.

Roy, P. (2008) 'Groundwater depletion hampers irrigation', *The Daily Star*, March 9, Dhaka.

Roy, R.C.K. (2000) *Land Rights of the Indigenous Peoples of the Chittagong Hill Tracts*, Copenhagen: International Working Group for Indigenous Affairs.

Rozario, S. (2006) 'The new burqa in Bangladesh: Empowerment or violation of women's rights?', *Women's Studies International Forum*, 29: 368–380.

Salway, S., Jesmin, S. and Rahman, S. (2005) 'Women's employment in urban Bangladesh: A challenge to gender identity?', *Development and Change*, 36(2): 317–349.

Save the Children (2006) *Watermarks: Child Protection during floods in Bangladesh*, Dhaka: Save the Children UK.

Schmuck, H. (2002) *Empowering Women in Bangladesh (Country Update)* (online). Available from: http://reliefweb.int/node/96333 [13 June 2011].

Schmuck-Widmann, H. (2001) *Facing the Jamuna River: Indigenous and engineering knowledge in Bangladesh*, Dhaka: Bangladesh Resource Centre for Indigenous Knowledge.

Schuler, R., Hashemi, S.M. and Badal, S.H. (1998) 'Men's violence against women in rural Bangladesh: Undermined or exacerbated by microcredit programmes?', *Development in Practice*, 8(2): 148–157.

Schuler, S. and Islam, F. (2008) 'Women's acceptance of intimate partner violence within marriage in rural Bangladesh', *Studies in Family Planning*, 39(1): 49–58.

Schuler, S., Bates, L. and Islam, F. (2008) 'Women's rights, domestic violence, and recourse seeking in rural Bangladesh', *Violence Against Women*, 14: 326–345.

Schuler, S.R. and Hossain, Z. (1998). 'Family planning clinics through women's eyes and voices: A case study from rural Bangladesh', *International Family Planning Perspectives*, 24(4): 170–175.

Schuler, S. R.,Yount, K. M. and Lenzi, R. (2012) 'Justification of wife beating in rural Bangladesh: A qualitative analysis of gender differences in responses to survey questions', *Violence Against Women*, 18(10): 1177–1191.

Schuler, S.R., Bates, L.M., Islam, F. and Islam, K. (2006) 'The timing of marriage and childbearing among rural families in Bangladesh: Choosing between competing risks', *Social Science & Medicine*, 62: 2826–2837.

Schuler, S.R., Hashemi, S.Y., Riley, A.P. and Akhter, S. (1996) 'Credit programs, patriarchy and men's violence against women in rural Bangladesh', *Social Science and Medicine*, 43(12): 1729–1742.

Seager, J. (2006) 'Noticing gender (or not) in disasters', *Geoforum*, 37: 2–3.

Seager, J. (1996) 'Hysterical housewives and other mad women', *Feminist Political Ecology: Global issues and local experiences*, D. Rochelaeu, Barbara P. Thomas-Slayter and Esther Wangari (eds), London: Routledge, pp. 271–284.

Selvaraju, R., Subbiah, A.R., Baas, S., and Juergens, I. (2006) 'Livelihood Adaptation to Climate Variability and Change in Drought Prone Areas of Bangladesh: Developing institutions and options,' *Asian Disaster Preparedness Centre*, Rome: Food and Agriculture Organization of the United Nations.

Semenza, J.C., Ploubidis, G.B. and George, L.A. (2011) 'Climate change and climate variability: Personal motivation for adaptation and mitigation', *Environmental Health*, 10(46) (Epub ahead of print).

Shehabuddin, E. (2008) 'Jamaat-i-Islami in Bangladesh: Women, democracy and the transformation of Islamist politics', *Modern Asian Studies*, 42(2/3): 577–603.

Shehabuddin, E. (1999) 'Contesting the illicit: Gender and the politics of fatwas in Bangladesh', *Signs: Journal of Women in Culture and Society*, 24(4): 1011–1044.

Shildrick, M. (1997) *Leaky Bodies and Boundaries: Feminism, postmodernism and (bio)ethics*. London: Routledge.

Shiva, V. (2005) 'Epilogue: Beyond reductionism', *Biopolitics; A Feminist and Ecological Reader on Biotechnology*, Vandana Shiva and Ingunn Moser (eds), London: Zed Books.

Shiva, V. (1993) 'The Chipko women's concept of freedom', *Ecofeminism*, Maria Mies and Vandana Shiva (eds), London: Zed Books.

Shiva, V. (1989) *Staying Alive: Women, ecology and development*, London: Zed Books.

Shoji, M. (2010) 'Does contingent repayment in microfinance help the poor duringnatural disasters?', *Journal of Development Studies*, 46(2): 191–210.

Shucksmith, M. (2004) 'Young people and social exclusion in rural areas', *Sociologia Ruralis*, 44(1): 43–59.

Siddiqi, D.M. (2003) 'The Sexual Harassment of Industrial Workers: Strategies for intervention in the workplace and beyond.' CDP-UNFPA Publication Series, Paper 26, Dhaka: Centre for Policy Dialogue.

Silverman, J.G., Gupta, J., Decker, M.R., Kapur, N. and Raj, A. (2007) 'Intimate partner violence and unwanted pregnancy, miscarriage, induced abortion, and stillbirth among a national sample of Bangladeshi women', *International Journal of Obstetrics and Gynaecology*, 114: 1246–1252.

Simmons, R. (1996) 'Women's lives in transition: A qualitative analysis of the fertility decline in Bangladesh', *Studies in Family Planning*, 27(5): 251–268.

Singh, S., Cabigon, J.V., Hossain, A., Kamal, H. and Perez, A.E. (1997) 'Estimating the level of abortion in the Philippines and Bangladesh', *International Family Planning Perspectives*, 23(3): 100–144.

Smith, D. (1992) 'Feminist reflections on political Eeconomy', *Feminism in Action*, M. Patricia Connelly and Pat Armstrong (eds), Toronto: Canadian Scholars Press, pp. 1–21.

Smith, L. (2006) 'Stern's report: "If we act now, we can avoid the very worst"', *The Times* 31 October, available from: www.timesonline.co.uk/tol/news/uk/article619828.ece [9 September 2010].

Snyder, M. (2006) 'Unlikely godmother: The UN and the global women's movement', *Global Feminism: Transnational women's activism, organising and human rights*, Myra Marx Ferree and Aili Mari Tripp (eds), New York: New York University Press, pp. 24–50.

South Asian Association for Regional Cooperation (2008) *SAARC Statistical Yearbook 2008* [online]. Available from: www.bbs.gov.ba [1 June 2011].

Stern, N. (2006) *Report of the Stern Review: The economics of climate change*, London: HM Treasury.

Stern, N. (2006) *The Economics of Climate Change: The Stern review* Cambridge Cambridge: University Press.

Sudo, N., Sekiyama, M., Watanabe, C., Bokul, M.H. and Ohtsuka, R. (2004). 'Gender differences in food and energy intake among adult villagers in northwestern Bangladesh: A food frequency questionnaire survey', *Int J Food Sci Nutr*, 55(6): 499–509.

Sultana, A.M., Jawan, J.A. and Hashim, I. (2009). 'Influence of purdah (veil) on education and employment of women in rural communities', *European Journal of Social Sciences*, 11(2): 267–280.

Suran, L., Amin, S., Huq, L. and Chowdhury, K. (2004) 'Does dowry improve life for brides? A test of the bequest theory of dowry in rural Bangladesh', *Population Council Policy Research Division Working Papers No.195* [online]. Available from: www.popcouncil.org/pdfs/wp/195.pdf [4 June 2011].

Tanner T.M., Hassan A., Islam K.M.N., Conway, D., Mechler R., Ahmed A.U., and Alam, M. (2007) *ORCHID: Piloting climate risk screening in DFID Bangladesh.* Detailed research report. University of Sussex: Institute of Development Studies.

The Economist (2013) 'Who owns Bangladesh? Women and property rights. Available from: www.economist.com/blogs/banyan/2013/08/women-and-property-rights [7 December 2013].

Thomalla, F., Cannon, T., Huq, S., Klein, R. and Schaerer, C. (2005) 'Mainstreaming adaptation to climate change in coastal Bangladesh by building civilsociety alliances', *Proceedings of the Solutions to Coastal Disasters 2005 Conference*, (May 8–11, 2005, Charleston, SC), Louise Wallendorf, Lesley Ewing, Spencer Rogers and Chris Jones (eds). Available from: www.pik-potsdam.de/research/research-domains/transdisciplinary concepts-and-methods/favaia/pubs/thomalla_etal_2005.pdf [15 July 2011].

Tovar-Restrepo, M. (2010). 'Climate change and indigenous women in Colombia', *Gender and Climate Change: An introduction*, Irene Dankelman (ed.), London: Earthscan, pp. 145–151.

Tschakert, P. (2011). Presentation to the Limits to Adaptation Working Group, Penn State University team, May 17.

Tschakert, P., Ziervogel, G., Sallu, S., Shackleton, S., Alston, M. and Koelle, B. (2011) *Barriers and Limits to Climate Change Adaptation: A conceptual framework.* Enquiries tomargaret.alston@monash.edu.

Tuhiwai Smith, L. (1999) *Decolonising Methodologies: Research and indigenous peoples.* London: Zed Books.

UNDP (2010) *Hum Dev Report 2010: The wealth of nations: Pathways tohum dev* [online]. Available from: http://hdr.undp.org/en/media/HDR_2010_EN_Complete_reprint.pdf [3 June 2011].

UNDP (2004) *A Global Report: Reducing disaster risk: A challenge for development*, available from: www.undp.org/bcpr.

United Nations Framework Convention on Climate Change (UNFCCC) (2011). *Article 1, Definitions.* Available from: http://unfccc.int/essential_background/convention/background/items/2536.php [9 September 2014].

UNHCR (2005) *The World's Women* 2005*: Progress instatistics.* Available from: http://unstats.un.org/unsd/demographic/products/indwm/wwpub.htm [27 November 2012].

UNICEF (2013) *The Millennium Development Goals*, available from: www.unicef.org/mdg/gender.html [15 January 2013].

UNICEF (2010) *Women and Girls in Bangladesh*, Dhaka: UNICEF Bangladesh.

UNICEF (2009) *Situation Assessment and Analysis of Children and Women in Bangladesh*, Dhaka: UNICEF.

UNICEF and BBS (2010) *Multiple Indicator Cluster Survey (MICS) 2009*, Bangladesh.

UN International Strategy for Disaster Reduction (2007) *Gender Perspective: Working together for disaster risk reduction; Good practices and lessons learned* – Examples of good practice, Geneva. (Included in annex of resources).

United Nations (UN) (2005) *Poverty*, available from: www.un.org/cyberschoolbus/briefing/poverty/poverty.pdf [4 July 2013].

United Nations (UN) (1995) *Beijing Platform for Action*, www.un.org/esa/gopher-data/conf/fwcw/off/a–20.en [23 August 2013].

United Nations Environment Programme (UNEP) (2005) 'Feature focus: Gender, poverty and environment', *GEO Yearbook 2004/5*. Nairobi: UNEP.

United Nations Human Rights Council (UNHRC) (2010) Joint Report of the Independent Expert on the Question of Human Rights and Extreme Poverty, Magdalena Sepulveda Cardona, and the Independent Expert on the Issue of Human Rights Obligations Related to Access to Safe Drinking Water and Sanitation, Catarina de Albuqerque, available from: http://daccess-dds-ny.un.org/doc/UNDOC/GEN/G10/158/79/PDF/G1015879.pdf?OpenElement [9 August 2013].

United Nations News Centre (2012) 'Doha meeting must take decisive action to tackle growing crisis of climate change – Ban', available from: www.un.org/apps/news/story.asp?NewsID=43671&Cr=climate+change&Cr1= [6 December 2012].

United Nations Secretary-General's High-level Panel on Global Sustainability (2012) *Resilient People, Resilient Planet: A future worth choosing*, New York: United Nations, available from: www.un.org/gsp/sites/default/files/attachments/GSPReport_unformatted_30Jan.pd [5 February 2012].

UN WomenWatch (2011) 'Women, gender equality and climate change', available from: www.un.org/womenwatch/feature/climate_change/downloads/Women_and_Climate_Change_Factsheet.pdf [29 November 2011].

Vidal, J. (2013) 'China and India "water grab" dams put ecology of Himalayas in danger', *The Observer*, August 10, available from: www.theguardian.com/global-development/2013/aug/10/china-india-water-grab-dams-himalayas-danger [6 September 2013].

van Schendel, W. (2009) *A History of Bangladesh*, Cambridge: Cambridge University Press.

Vincent, K., Cull, T. and Archer, E. R.M. (2010) 'Gendered vulnerability to climate change in Limpopo Province, South Africa', *Gender and Climate Change: An introduction*, Irene Dankelman (ed.), London: Earthscan pp. 130–137.

Women's Environment and Development Organisation (WEDO) (2008) *Gender, Climate Change and Human Security: Lessons from Bangladesh, Ghana and Senegal*, New York: WEDO.

World Bank (2013) 'Physicians (per 1,000 people)', available from: http://data.worldbank.org/indicator/SH.MED.PHYS.ZS [25 July 2013].

World Bank (2010) *Country Assistance Strategy for the People's Republic of Bangladesh Report No. 54615-BD* available from: http://documents.worldbank.org/curated/en/2010/07/12649110/bangladesh-country-assistance-strategy-period-fy11-14 [9 September 2014].

World Bank (2009) 'Public attitudes toward climate change: Findings from a multi-country poll', *World Development Report 2010: Development and climate change*, Washington, DC: World Bank.

World Bank (2008) 'Education for all in Bangladesh. Where does Bangladesh stand in achieving the EFA goals by 2015?', *Bangladesh Development Series, Paper No.24*, Dhaka: World Bank.

World Bank (2006) 'Bangladesh country environmental analysis', *Bangladesh Development Series, Paper No.12*, Dhaka: World Bank.

World Bank (2000) Bangladesh Climate Change and Sustainable Development, Washington, DC: World Bank.

World Health Organization (2012) 'Water global analysis and assessment of sanitation and drinking water', *GLAAS 212 Report UN*, WHO publishing, available from: www.un.org/waterforlifedecade/pdf/glaas_report_2012_eng.pdf [14 December 2013].

Yang, L. (2011) 'Bangladesh adopts new policy for uplifting women's status', 3 August, available from: http://news.xinhuanet.com/english2010/world/2011-03/08/c_13767164.htm [21 August 2013].

Zalewski, M. (2010) '"I don't even know what gender is": A discussion of the connections between gender, gender mainstreaming and feminist theory', *Review of International Studies*, 36: 3–27.

Zaman, H. (2001) 'Paid work and socio-political consciousness of garment workers in Bangladesh', *Journal of Contemporary Asia*, 31(2): 145–160.

Zaman, H. (2000) *Assessing the Poverty and Vulnerability Impact of Micro-credit in Bangladesh: A case study of BRAC*, Washington, DC: World Bank.

Zaman, H. (1999) 'Violence against women in Bangladesh: Issues and responses', *Women's Studies International Forum*, 22(1): 37–48.

Zaman, H. (1996) *Women and Work in a Bangladeshi Village*, Dhaka: Narigrantha Prabartana.

Zaman, H. (1995) 'Resistance against Seclusion: Women, work and patriarchy in Bangladesh', *Canadian Journal of Development Studies*, pp. 105–122.

Zafarulla, H. (2000) 'Through the brick wall, and the glass ceiling: Women in the civil service in Bangladesh', *Gender, Work and Organisation*, 7(3): 197–120.

Zelezny, L.C., Chua, P. and Aldrich, C. (2000) 'Elaborating on gender differences in environmentalism', *Journal of Social Issues*, 56: 443–457.

Index

For Product Safety Concerns and Information please contact our EU representative GPSR@taylorandfrancis.com
Taylor & Francis Verlag GmbH, Kaufingerstraße 24, 80331 München, Germany

www.ingramcontent.com/pod-product-compliance
Ingram Content Group UK Ltd.
Pitfield, Milton Keynes, MK11 3LW, UK
UKHW022129190926
475702UK00013B/1714
9780815361732